OPERATION, CARE, AND REPAIR
OF
FARM MACHINERY

Printed in Canada

Originally published by The John Deere Company, c. 1937

10   9   8   7   6   5   4   3   2   1

The Library of Congress Cataloging-in-Publication Data
is available on file.

Operation, Care, and Repair

. . . of . . .

Farm Machinery

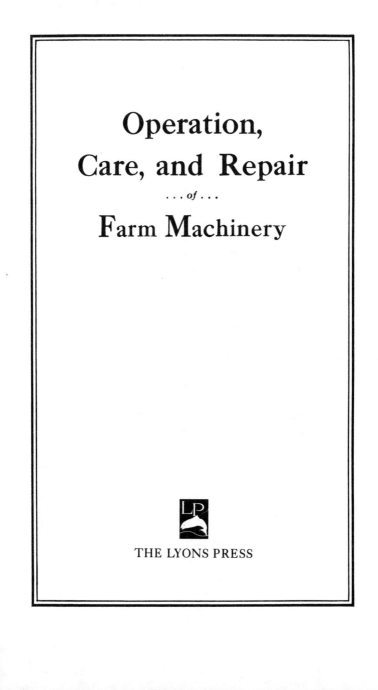

THE LYONS PRESS

Contents

Preface

THE study of the operation, care, and repair of farm implements becomes increasingly important with the steady growth in the use of farm machines. The farmer of today, with his machines of greater working capacities, must be a more efficient mechanic than were his father and his grandfather, whose machines were comparatively simple.

The high schools and agricultural colleges of the United States are doing a very commendable work in teaching farm mechanics to the young men who will be the farmers and agricultural leaders of tomorrow. The activity of these institutions in encouraging the use of more efficient machines and in teaching the proper methods of adjustment and operation of all farm implements, is certain to result in the lowering of production costs throughout the agricultural industry.

It is the purpose of this book to assist instructors in farm mechanics in giving a thorough and practical course in the operation, care, and repair of the more important farm machines. It is designed especially for those who make a study of farm mechanics in high schools, colleges, and short courses with the expectation of applying the knowledge gained to actual farming or to the instruction of others.

The unqualified endorsement given the first editions of this book by educators the country over led to the production of this, the eleventh edition. It is the hope of the publisher that this more thorough and complete textbook will be of even greater value than its predecessors in filling the need for a complete text dealing only with the operation, care, and repair of farm machines.

The publisher wishes to express appreciation to high school instructors in agriculture, agricultural college professors, and state supervisors of agricultural education for the assistance and suggestions so generously given in the preparation of this and the previous editions.

JOHN DEERE

Part One

PREPARATION OF THE SEED BED

BEFORE discussing the implements used in preparing the soil for planting, it is well to consider, briefly, the results sought by their use. A clear understanding of the purpose and value of the machine studied is necessary before details of its operation and care can be fully appreciated. Thus, the reasons for certain adjustments on a plow become quite evident when we know that the purpose of plowing is to pulverize the soil and cover field trash.

The Ideal Seed Bed. It is unwise to say that any one type of seed bed is desirable for all crops and all soils. The make-up and process of preparing what might be called a good seed bed in gumbo soil would be greatly different from that used in a sandy soil.

In general, the seed bed should be roomy, thoroughly pulverized, and compact. It should have perfect contact with the subsoil to facilitate the rise of moisture. Large air spaces, bunches of field trash, and hard lumps or clods are undesirable, as their presence retards root growth and breaks the contact with the subsoil. The operations used in preparing a seed bed will vary with soil and field conditions and the results wanted.

The Plow. The plow is the most important implement used in seed bed preparation. Its purpose is to pulverize or break up the soil, admitting air and light—two essentials to normal plant growth. The plow covers surface trash or manure, mixing it with the soil to decay and furnish plant food.

There are many different types of plows, but all serve the same purpose. The listing plow, the middlebreaker, the disk

1

plow, and regular moldboard plow are all used in stirring and pulverizing the soil preparatory to planting.

The Disk Harrow. Next to the plow in importance comes the disk harrow. It is a valuable implement when used before and after plowing, or when used alone in preparing seed beds for some crops. It pulverizes clods, mixes trash with the soil, and forms a mulch when used before plowing. When used after plowing, it chops lumps, closes air spaces, and makes the seed bed compact.

Spike- and Spring-Tooth Harrows. Finishing the seed bed and destroying weeds before and after planting are the main purposes of spike- and spring-tooth harrows. Their crushing and stirring effect breaks up clods and crusted top soil, leaving a fine surface mulch for planting or for plant growth.

Soil Pulverizer or Packer. The pulverizer, or packer, as it is more commonly called, crushes lumps, closes air spaces, and leaves the seed bed firm, in ideal condition for planting. The packer is used extensively in territories where the soil tends to blow. It leaves an irregular, firm surface which does not blow as readily as a looser, more regular surface.

Wheatland Implements. In districts where soil conditions are suitable and winter wheat is raised on a large scale, the problem of preparing the seed bed is one of speed and low cost. Here, the disk tiller, listing plow, ridge burster, and, in the semi-arid regions the damming lister, meet these requirements very satisfactorily.

Chapter I.
PLOWS
Plow Bottoms

Importance of the Bottoms. A plow is no better than its bottoms. No matter how well the frame may be built, how strong its beams, how modern its design, the plow will be only as satisfactory as its bottoms. If the bottoms fail to scour and turn the soil properly, the seed bed will be uneven and lumpy, resulting in lower yields. If the bottoms turn an even furrow, cover trash well, and pulverize the furrow slice as desired, a uniform seed bed will result.

Costly delays at plowing time are often caused by plow bottoms that refuse to scour. The trouble may be in the way the bottoms are made. It may be in the adjustment of the hitch, it may be due to dull or improperly set shares, or to looseness or misalignment of the bottoms on the beams. The plowman must be constantly on the alert for signs of inefficiency in his plow bottoms.

Parts of the Bottom. The plow bottom consists of share, landside, moldboard, and frog (Fig. 1). The share and landside act as a wedge in the soil, cutting the furrow loose from the sub-soil much as a wedge splits a log. The curved surface of the upper part of the share and

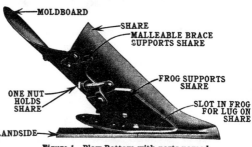

Figure 1—Plow Bottom with parts named.

the properly curved moldboard act as a single curve to invert the furrow slice. In passing over this curved surface, the furrow is twisted and broken, and the soil is pulverized, mixed, and aerated.

The steel frog holds the bottom parts and beam together. The landside and moldboard are bolted solidly to the frog. Most riding plows and tractor plows of the moldboard type now have quick-detachable shares. To remove or replace the share, it is necessary to loosen but one nut (Fig. 1). This quick-detachable feature saves time when shares are removed for sharpening.

Certain types of chilled-iron plows have a detachable

chilled shin-piece that serves as a long-wearing cutting edge for the shin of the moldboard where the hardest wear occurs (Fig. 2).

Types of Bottoms. Different types of soils require different shapes of bottoms to accomplish the results desired in

plowing. The texture of the soil and the amount of moisture it contains determine whether it should be thoroughly pulverized or merely turned over, to be pulverized with other implements. A mellow loam soil and soils

Figure 2—Chilled-iron bottom with detachable shin-piece.

of similar texture should be plowed with a bottom that will pulverize well, while a sticky, wet clay soil should be plowed with a bottom that will break it as little as possible, leaving the pulverizing to be done with other machines.

The pulverizing effect of a plow depends upon the shape of its bottom. A bottom with a long, gradual curve in the moldboard turns the furrow slice gently and disturbs its composition but little. The other extreme is the short, abruptly-curved moldboard that twists and shears the soil as it passes over, making a mellow, well-pulverized furrow. The pulverizing effect produced by the curved surface of the moldboard is illustrated in Fig. 3.

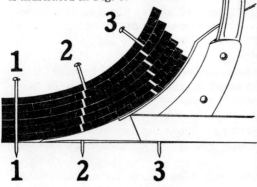

Figure 3—Principle of the pulverizing effect of the plow bottom under most soil conditions. The shearing effect produced by the curved surface is illustrated by pins 1, 2, and 3. Pin 1 is sheared into many parts when it reaches position of pin 3. Bending the pages at the corner of a book will illustrate this principle. The breaking effect produced by the curved surface is also illustrated.

Between these two extremes are many types of bottoms designed to meet many different soil conditions, but for general use bottoms may be classified as breaker, stubble, general-purpose, slat moldboard, and black-land. The breaker (Fig. 4) is used in tough sod where complete turning of the furrow slice without materially disturbing its texture is desired. Stubble bottoms (Fig. 5) are especially adapted to plowing in old ground where good pulverizing of the soil is desired. General purpose bottoms (Fig. 6) meet the demand for bottoms that will do good work in stubble, tame sod, old ground, and a variety of similar conditions. The general purpose is designed to do satisfactory work in the varying conditions found on the average farm.

Figure 4—Prairie breaker.

Figure 5—Stubble bottom.

Figure 6—General purpose bottom

The slat moldboard bottom (Fig. 7) is used in loose, sticky soils, and the black-land bottom (Fig. 8) in gumbo and "buckshot" soils. In both types of soil, scouring is a serious problem.

The deep tillage bottom (Fig. 9) is used in certain restricted territories where it is desirable to plow to unusually great depths—as deep as sixteen inches.

There are a number of variations of these general bottom shapes built to meet a wide variety of soil conditions but, in every case, the manufacturer pro-

Figure 7—Slat moldboard bottom.

Figure 8—Black-land bottom.

Figure 9—Deep tillage bottom.

vides the implement dealer with the types of bottoms suited to his territory.

Materials Used in Bottoms. Classified according to materials used in manufacture, there are two kinds of plow bottoms—steel and chilled cast-iron. Steel bottoms may be either solid steel or hardened soft-center steel. The latter bottoms are in more general use. In some sections, the soil conditions are such that a combination of chilled-iron shares and soft-center steel moldboards is used with excellent results. Both steel and chilled bottoms are used by some farmers who have varying soil conditions on their farms.

Soft-Center Steel Bottoms. A very highly polished fine-textured steel moldboard is necessary to good scouring in sticky, fine-grained soils. Soft-center steel has the necessary hardness and thickness for good scouring and long wear on the outer surfaces, and strength enough in the inner layer to withstand shocks and heavy loads in difficult soils. The outside layers are very high in carbon, extremely hard, dense steel.

Figure 10—Genuine soft-center steel. Nos. 1 and 3 are layers of high-carbon steel, harder than what is known as "tool steel." No. 2 is a layer of soft, tough steel, the hard steel having been fused around it. Note uniform thickness of layers.

Figure 11—Genuine soft-center steel share point (landside toward figures). A—Patch of hard tool steel. 1 and 3—Hard steel. 2—Soft steel. 4—Steel landside, lap-welded. Note thickness of hard-steel layers and how lap-weld stiffens point.

Between these two hard layers is a layer of soft, tough steel, the hard steel having been fused to it (Fig. 10). In genuine soft-center steel, all three layers are uniformly thick. There is no out-cropping of soft spots, no thin places in the outer layers to wear through rapidly. Fig. 11 shows cross section of genuine soft-center steel share point, illustrating how it is reinforced by lap-welding.

Solid Steel Bottoms. Solid steel bottoms are used in soils where scouring as a rule is not difficult. They are made of solid steel and are not tempered. They should not be used in sandy or gravelly soil, as they would tend to wear too rapidly. Solid steel shares are sometimes used with soft-center steel moldboards where soil conditions do not require the more costly soft-center steel shares for scouring.

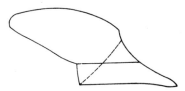

Figure 12—Plow bottom showing area, below dotted line, which receives 75 per cent of the draft when plowing, illustrating the necessity of keeping shares sharp for light draft of the plow.

Chilled-Iron Bottoms. Plow bottoms made of chilled iron are designed primarily for use in sandy or gravelly soil where the share and moldboard must withstand the scratching and hard wear of a soil of this type, and where the denser, finer-grained surface of soft-center steel is not necessary for scouring. The

Figure 13—Heavy lines show proper shape of sharp share points for good penetration. Dotted lines show how worn points look before sharpening.

material used in these bottoms is extremely hard and long wearing due to a process called "chilling."

In casting chilled shares, a piece of metal called a "chill" is placed into the mold along the cutting edge and point where the finished share is to be chilled. When the hot metal comes into contact with the "chill," the sudden cooling leaves the grain of the metal at right angles to the surface. Thus, the dirt rubs the ends of the grain in the metal when passing over the share. A smooth and long-wearing surface results. Chilled shares may be sharpened by grinding but, because of their low cost, it is usually more satisfactory to replace worn shares with new ones.

Sharpening Plow Shares. The share is the most vital part of the bottom. It is the "business end"—the pioneer part in all of the work that a plow does. Draft, penetration, steady-running, and good work all depend upon the share.

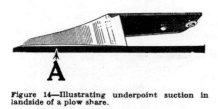

Figure 14—Illustrating underpoint suction in landside of a plow share.

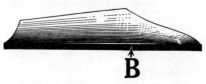

Figure 15—Illustrating underpoint suction in throat of a plow share.

Note Fig. 12, which shows the area of the plow bottom that is responsible for 75 per cent of the draft when the plow is at work. This illustration clearly shows the importance of keeping the share in good cutting condition at all times. Note, also, Fig. 13. A dull share may cause poor penetration and may greatly increase the draft of a plow. A sharp, correctly set share adds to the efficiency and good work of the plow bottom.

Many farmers have shop equipment for sharpening their plow shares, while the great majority depend upon local blacksmiths or mechanics for this service. In either case, it is well to know how shares should be sharpened.

When sharpening soft-center or solid steel shares, the point of the share should be heated to a low, cherry red (not too hot) and hammered on the top side until the point is sharp. Hammering should be done at a cherry red only. Working the share at a high heat destroys the quality of the steel. The entire cutting edge should be drawn from the underside until sharp. Only as much as can be hammered should be heated at one time. The body of the share should not be heated while sharpening, but should remain cool to prevent warping and disturbing of the fitted edges.

Should the share get out of shape or the fitted edges become warped during the sharpening process, the entire blade should be restored to proper shape before hardening. This can be done best at a black heat.

Soft-center steel shares should be hardened after sharpening. To do a thorough job of hardening, it is necessary to prepare the fire to heat the entire share uniformly to a cherry red. Care should be used in getting the heat uniform. The share is taken from the fire and dipped into a tub of clean, cold water with the cutting edge down. Care should be taken to keep the blade perpendicular during this process.

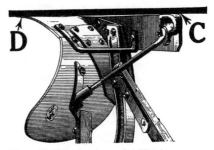

Figure 16—Wing bearing, point "D", is necessary to smooth running of walking plows.

Solid steel shares should not be hardened.

Setting Shares for Suction. The plow bottom is led into the

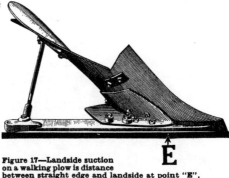

Figure 17—Landside suction on a walking plow is distance between straight edge and landside at point "E".

ground and held to its work by the underpoint suction of the share. Such suction is produced by turning the point of the share down slightly below the level of the underside of the share. (See Fig. 14.) The amount of suction necessary depends upon the type of plow and existing soil conditions. Stiff clay soils are harder to penetrate than light loam soils and require more suction in the share point.

Landside suction (see Fig. 17) in a plow share holds the bottom to its full-width cut. It is produced by turning the point of the share toward the unplowed ground. The land suction, as well as the down suction, should be measured when the share is new, so that the same amount of suck can be given the share when it is sharpened.

The importance of having the correct amount of suck in the share cannot be emphasized too strongly. Too little under-point suction will cause the plow to "ride out" of the ground and cut a furrow of uneven depth. Too much will cause "bobbing" and heavy draft. In both cases, the plow is difficult to handle. If the landside suction is too great, the bottom tends to cut a wider furrow than can be handled properly, and the reverse is true when the landside suction is not sufficient.

The setting of walking plow shares and of riding and tractor plow shares is discussed separately, as the shares are different.

Directions for setting walking plow shares are as follows: Set the point of the share down so there is 1/16- to 1/8-inch

Figure 18—Bottoms are the "business end" of the plow, for no plow is better than its bottoms.

suction, or clearance, under landside at point "A" (Fig. 14). The clearance, or underpoint suction in the throat of the share should be 1/16- to 1/8-inch at point "B" (Fig. 15). All 12-, 14-, and 16-inch walking plow shares should have a wing bearing. The correct wing bearing (point "D", Fig. 16) is as follows: 16-inch plow, 1-1/2 inches; 14-inch plow, 1-1/4 inches; 12-inch plow, 3/4-inch. A straight-edge placed at rear of the landside (point "C") and extending to wing of share should touch back of edge (point "D", Fig. 16). When sharpening the share, care must be taken not to turn point to one side or the other. When fitted to the plow, there should be about 1/4-inch clearance or landside suction at "E" (Fig. 17).

For riding and tractor plows, set the point of the share down until there is 1/8- to 3/16-inch suction under the landside at point "A" (Fig. 14). See that clearance in throat of share at "B" (Fig. 15) is at least 1/8 inch. Set edge of share at wing point, "F", without wing

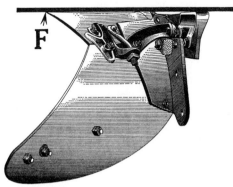

Figure 19—Riding and tractor plows do not require wing bearing at "F".

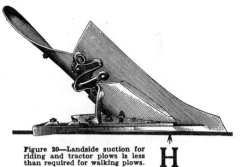

Figure 20—Landside suction for riding and tractor plows is less than required for walking plows.

bearing (Fig. 19). For landside, set should be about 3/16-inch clearance at "H" (Fig. 20).

Care of the Bottom. The plow bottom will give the best satisfaction when given the best care. If kept in good condition, it will give little scouring trouble. If permitted to rust, it may cause any amount of hard work and lost time.

One of the first rules a plowman should learn is to polish the bright surfaces of his plow bottoms and apply a light coating of oil whenever the plow is not in use. Strict observance of this rule will save many hours of difficulty in getting a rusted surface repolished. A heavy coating of a good hard oil should be applied to the bottoms when storing the plow from season to season.

Plow manufacturers paint or varnish the surfaces of new plow bottoms to protect them from moisture from factory to user. This protective coating must be removed before the plow is taken into the field. This can be accomplished best by means of a paint and varnish remover which is obtainable at most paint stores. A can of concentrated lye dissolved in two or three quarts of water will serve the same purpose. The solution should be applied with a swab or a piece of gunny sack, the operator being careful not to get it on his hands. After the coating has been softened in this manner, it can be scraped off readily with a putty knife or similar instrument, care being taken not to scratch the polished surface.

If the new plow is not to be used immediately after the protective covering has been removed, the bottoms should be oiled, as the metal rusts readily if exposed to the air after treatment with the suggested strong solutions.

In case a plow bottom becomes badly rusted, working it in a coarse sandy or gravelly soil will aid in restoring a land polish.

Rolling Coulter and Jointer. One of the most important duties of the plow is to cover the stubble, stalks, or other trash usually found on the surface of a field. Thorough covering of such matter hurries its decomposition and makes cultivation of future crops less difficult than when trash is left on top of the seed bed to clog cultivating machines.

The rolling coulter and jointer attachments for moldboard plows have proved to be big aids to clean plowing and good covering. The rolling coulter cuts through the surface trash and aids in securing a clean furrow wall, reducing the draft on the cutting edge of the plow bottom. In reality, the jointer is a miniature plow, the purpose of which is to cut a small furrow off the main furrow slice and throw it toward the furrow in such a manner that all stubble and trash are buried in the bottom of the furrow.

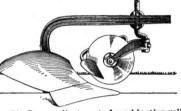

Figure 21—Proper adjustment of combination rolling coulter and jointer for ordinary plowing conditions.

To get the best results with the combination rolling coulter and jointer, shown in Fig. 21, the hub of the coulter should be set about one inch back of the share point with the blade running just deep enough to cut the trash, about three to four inches in ordinary conditions. The jointer should cut about two inches deep. There should be about 1/8 inch space between the jointer and the coulter blade.

The independent rolling coulter and jointer, which permit a wider range of adjustment to meet field conditions, are gaining in popular use.

Figure 22—When used alone, the coulter should be set 1/2- to 5/8-inch to the land, and the hub should be about 3 inches to the rear of share point.

When the rolling coulter is used alone, it should be set about 1/2 to 5/8 inch to the land (Fig. 22). The hub of the coulter should be about three inches behind the point of the

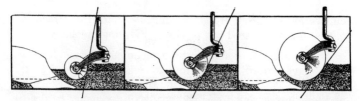

Figure 23—Large rolling coulters mount trash more readily than small ones, thereby insuring a clean-cut furrow with less draft.

share.   In soil that does not scour well, more pressure on the moldboard can be secured by setting the coulter farther to land.   If there is considerable trash on the field, the coulter should be set just deep enough to cut it—if set too deep, it pushes instead of cuts trash. The larger rolling coulters prove more effective in trashy conditions, as they mount trash more effectively than smaller coulters.   (See Fig. 23.)   When plowing sod, the coulter must be deep enough to cut the roots below the surface, usually about one inch shallower than the share is cutting.

Keeping the rolling coulter sharp and well oiled, and the jointer sharp and properly set will add greatly to the efficiency of their work.

Questions

1.   *Why is a good bottom essential to good plowing?*
2.   *Name and tell the purpose of the parts of a plow bottom.*
3.   *How is the plow share removed for sharpening?*
4.   *Describe the relation between shape of moldboard and its pulverizing effect upon the soil.*
5.   *Name the five general types of bottoms and tell purposes of each.*
6.   *In what kinds of soils are chilled cast-iron bottoms used?*
7.   *What is the difference between solid-steel bottoms and soft-center steel bottoms?   Tell purpose of each.*
8.   *Why is it important to keep plow shares sharp?   Describe process of sharpening soft-center and solid steel shares.*
9.   *What is meant by "underpoint suction" and "landside suction" in a plow share, and what is the purpose of each?   Describe results of improper suction and tell how to measure correct set for riding and walking plows.*
10.   *How would you care for plow bottoms when in use and when not in use?   How would you remove factory varnish from plow bottoms?*
11.   *Describe purpose and proper adjustment of the rolling coulter and jointer.*

Walking Plows

Distribution and Use. On the great majority of farms, the walking plow has been supplanted by riding and tractor plows for general field work. However, in some sections, the walking plow is still depended upon to handle the bulk of the plowing. In certain cotton-raising districts in the South, and in small farm districts all over the United States, walking plows are used almost exclusively. Every farm, large or small, needs a walking plow for plowing gardens and small plots, but the general trend is to plows of larger capacity.

Steel and chilled cast-iron walking plows are built in a great variety of styles to meet a wide variety of soil conditions. The discussion of plow bottoms on a preceding page furnishes a picture of the requirements and the way they are met by plow manufacturers. Fig. 24 illustrates and names the parts of a general-purpose walking plow. Other types of walking plows are similarly constructed.

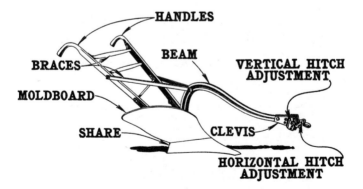

Figure 24—Common type of walking plow with important parts named.

The hillside or swivel plow is a variation of the walking plow designed to turn all furrows one way, plowing back and forth.

Figure 25—Walking middlebreaker.

Walking Middle-breaker. The middle-breaker (see Fig. 25) was developed to meet the request of the southern cotton grower for a double-moldboard plow that would turn a furrow each way, burst out a ridge, or bed up a new row for planting, in one operation. The middlebreaker reduced production costs in these regions, and its principle has since been adapted to the lister and the listing plow.

Hitching Walking Plows. The proper adjustment of the hitch is the most important factor in the operation of a plow. The kind of work a walking plow will do, its draft, and its

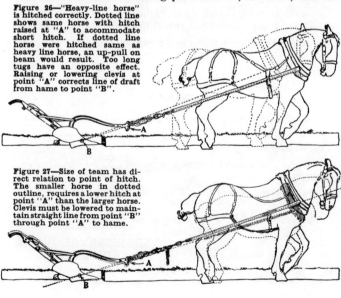

Figure 26—"Heavy-line horse" is hitched correctly. Dotted line shows same horse with hitch raised at "A" to accommodate short hitch. If dotted line horse were hitched same as heavy line horse, an up-pull on beam would result. Too long tugs have an opposite effect. Raising or lowering clevis at point "A" corrects line of draft from hame to point "B".

Figure 27—Size of team has direct relation to point of hitch. The smaller horse in dotted outline, requires a lower hitch at point "A" than the larger horse. Clevis must be lowered to maintain straight line from point "B" through point "A" to hame.

handling qualities depend to a great extent upon the correct relation between power and load.

By following a few simple rules, the most inexperienced can adjust his plow to run right. The most important rule to observe is this: the point of hitch should be on a straight line drawn from center of load to the center of power when plow is at work.

Fig. 26 shows the effect of long and short tugs on the work of a plow and the adjustments necessary to get correct working position of the hitch. Fig. 27 illustrates the changes necessary in the vertical or up-and-down hitch to accommodate horses of different sizes. The correct vertical hitch of walking plows is necessary to smooth running, correct depth, and the comfort of man and beast.

The horizontal adjustment of walking-plow hitches is comparatively easy, provided the doubletree is about the right length and the share is properly set. If a right-hand plow is not taking enough land, the clevis may be moved one or two notches to the right; if taking too much land, an adjustment of one or two notches to the left will bring the desired results. A left-hand plow is adjusted in an exactly opposite manner.

Questions

1. *In what section of the country are walking plows used most extensively?*

2. *Name the parts of a walking plow.*

3. *How does a middlebreaker differ from the ordinary plow and for what is it used?*

4. *Why is proper hitching of great importance in operating a walking plow? What rule governs hitching?*

5. *What effect does the length of tugs have upon the hitching of walking plows?*

6. *How would you adjust vertical and horizontal hitches for best results?*

Riding Plows

Types and Distribution. The three general types of what are usually classed as riding plows are sulky plows, two-bottom gang, and three-bottom gang plows, each built in several different styles, to meet the requirements of a particular section of the country.

The sulky, or one-bottom riding plow, was the natural outgrowth of a desire for a plow more easily handled than the walking plow. The hard work of following and handling the walking plow in general field plowing, tested the endurance of the most willing farmer. The sulky increased his working capacity and lightened his task.

Then came the two-bottom gang which doubled the number of acres one man could plow in a day, using one or two horses more than used on a sulky.

The three-bottom gang came into popularity in big-farm districts as a result of the success of the two-bottom gang in lowering production costs.

Figure 28—Two-way plow, popular for plowing irrigated, hilly, or irregularly shaped fields.

Styles of Riding Plows. For general usage, there may be considered four styles of riding plows, each style meeting a particular need. One of these styles has a frame, while three are frameless—that is, all parts attach directly to the beam.

The three-wheel frame, or foot-lift, sulky is probably the most widely used style of the sulky plows. It is built much like the gang illustrated in Fig. 29, except that it has only one bottom. Directions for its operation and care are the same as those given for the gang plow in later paragraphs.

The frameless, or low-lift, is extremely simple and easy to handle. Three levers provide quick and accurate adjustment for good plowing.

The two-way plow, shown in Fig. 28, is especially adapted to plowing hilly fields, irregular fields, and irrigated fields that must be kept level for proper regulation of water flow in the ditches. The operator starts at one side of the field and plows back and forth until the field is finished, thus eliminating dead furrows and back ridges. There are two complete bottoms, one left-hand and one right-hand. Pressure on a pedal causes the power lift to raise the working bottom when the end of the field is reached. By means of a lever, the idle bottom can be put to work for the return trip down the field.

While the operation of each style of sulky plow is somewhat different from the operation of gang plows, to avoid confusion, the details of operation and adjustment of all

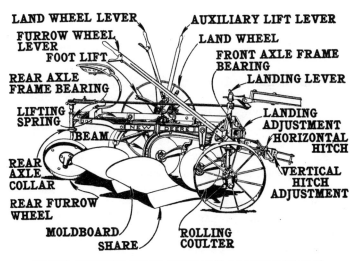

Figure 29—Two-bottom foot-lift gang plow with important parts named.

riding plows will be discussed collectively, using the two-bottom gang plow, shown in Fig. 29, as a basis.

Two- and Three-Bottom Gang Plows. The gang plows may be divided into frameless or low-lift, and frame or foot-lift styles. The frame styles are in more general use, although in certain limited sections the frameless styles are preferred.

The two-bottom frame or high-lift gang, shown in Fig. 29, is probably the most widely used style of two-bottom gang.

Care of Riding Plows. The importance of proper care for all types of plow bottoms has been emphasized in preceding paragraphs. The shares and coulters must be sharp and properly set, and the scouring surfaces of the entire bottom must be kept in good condition if the plow is to operate efficiently and satisfactorily.

Dust-proof and oil-tight boxings are provided on all wheels of riding plows (see Fig. 30). It is necessary that these boxings be kept generously greased with a good grade of hard oil to prevent excessive friction and wear. The regular oiling of rolling coulters will aid in their work and add years to their period of service. The liberal use of lubricants in the operation of plows as well as other farm machines will cut repair expenses and add to the life of the implements.

Adjustment and Operation of Riding Plows. When all parts of a riding plow are properly adjusted the combined weight of the plow, the driver, and the soil being turned is carried on the wheels. Heavy draft results when, because of wrong adjustments, all of the weight is not carried on the wheels, or wheels and rolling landside.

The first and most important point to remember is that shares must be sharp and properly set to do a good job of plowing. The entire bottom must be well polished for good scouring. Rolling coulters and jointers, if used, must be sharp and in correct position. The discussion of plow bottoms on preceding pages covers details of bottom and bottom-equipment adjustments.

Correct setting of the rear furrow wheel is necessary to uniform plowing and light draft. The wheel should run straight in the corner of the furrow on a level with the rear bottom. For proper penetration, there should be one-half inch space, or room enough to slip the fingers between heel of the landside and bottom of the furrow (Fig. 31). If there is less than this amount of space, the share points are raised

Figure 30—Wheel-boxing, showing dirt-proof and oil-tight construction. Collar on outer end of boxing holds wheel on axle and takes end thrust. Screw cap provides means of applying lubrication.

higher than landside heel, interfering with quick and easy penetration and causing plow to run shallower than desired in places, resulting in uneven plowing. If the clearance at heel of the landside is more than one-half inch, the share points run too deep and increase the draft of the plow. Moving up setscrew collar ("E" in Fig. 31) gives more space beneath landside heel. Moving it down has an opposite effect. There should also be one-half inch clearance

Figure 31—There should be enough clearance beneath heel of landside to permit fingers to pass as illustrated when plow is at work. Adjust at "E" for proper clearance.

or space enough to slip the fingers between the landside and furrow wall at heel of landside (Fig. 32) causing the furrow wheel to carry the landside pressure. Adjust bracket at "F" for correct setting.

The front furrow wheel should be set with a slight lead to the land. The lead should be just enough to hold the wheel in the corner of the furrow when the plow is in operation.

The ratchet landing device is attached directly to the front furrow wheel axle. With this, the plow can be given more or less land for work on hillsides or to straighten crooked furrows.

When the furrows, turned by a gang plow, do not lie alike, it is usually due to the furrow slices not being the same

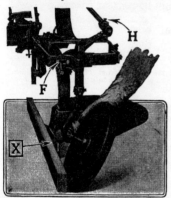

Figure 32—Showing test for clearance between landside heel and furrow wall to determine if furrow wheel is carrying landside pressure. If there is not room for the fingers between straight edge and landside (X), adjust at "F". Adjust at "H" on connection rod to set rear wheel to run straight.

width. This may be corrected by leveling the bottoms or adjusting the plow so both bottoms cut the same width furrows. The width of cut of the rear bottom is regulated b y adjusting the rolling coulters. The same applies to the front bottom unless the plow is old and worn, in which case it is necessary to set the front furrow wheel in, thereby narrowing cut of the front bottom. This is done by moving the landing adjustment casting in on the frame. (See Fig. 29.)

The foot lift on a riding plow acts both as a lift and a lock. The plow is locked down when in plowing position by pushing forward on the upper lift pedal until the lock goes over center. In plowing stony land, the setscrew on the lock can be screwed down to set the strap anti-lock. When so set, the lifting spring should be loosened sufficiently to prevent bottoms lifting excepting when they strike an obstruction. The bottoms will then maintain their depth, but will come up automatically when the shares strike an obstruction.

The lifting spring, which aids in raising the bottoms, should be adjusted with enough tension to make the plow lift easily. If left too loose, the plow will lift heavily.

Overhauling Riding Plows. The plow should be over-hauled and put in condition for the next season's work as soon as possible after finishing plowing. If new parts are needed before the plow can be used again, they should be ordered, attached, and the plow put in condition ready for the field, during slack seasons. The following instructions are designed to aid in a thorough inspection of riding plows:

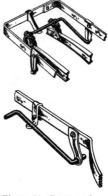

First—Examine the wheel box-ings. If they are badly worn, they should be renewed; if not, they should be slipped off the axle, and both axle and boxing washed clean with kerosene, and a fresh supply of grease applied. If the collar that holds the wheel to the axle is badly worn so as to allow excessive play of the wheel boxing on the axle, it should be replaced with a new one.

Second—Examine the shares. If they are not sharp, detach and have them sharpened and properly set. If they are worn excessively, get new

Figure 33—Front and side bail stops.

shares that are made by the manufacturer of the plow.

Third—On the high-lift, foot-lift plows, suspended by one or two bails, examine the bail stops. (See Fig. 33.) These are located on the front frame bar and on the right frame bar. They should be so adjusted that when the plow is locked down in plowing position, the bails rest securely on the stops. The bail bearings should also be examined. If they are worn loose, take off the cap and file or grind it until it fits snugly with all bolts tight. This will help greatly in keeping the plow running steadily and quietly.

Fourth—Examine rolling coulters and hub bearings. Coulters should be sharp and well polished. If the hub bearings are badly worn, replace them with new ones.

Fifth—Check up the location of the rear axle collar. (See Fig. 29.) This collar should support rear end of the frame and transmit weight of rider to the rear wheel. If it has become loose, permitting it to slip on the axle, weight of plow and rider will be carried on the bottom of the plow landside instead of on the rear wheel.

Sixth—The rear axle frame bearing carries the weight of the rider and plow and transmits the side pressure created by the moldboards to the rear furrow wheel. If the bearing becomes badly worn, the landside of the rear bottom will carry this pressure, resulting in heavy draft and excessive wear on the landside. On some makes of plows, this bearing is provided with a take-up casting with bolts at both upper and lower ends. This makes the proper adjustments simple and easy.

Seventh—The front furrow wheel axle bearing should be reasonably snug, in order to keep the wheel running at the proper angle, and the front furrow to proper width.

Eighth—Look over the entire plow for loose nuts and worn bolts. A plow operates at great disadvantage when parts are loose, due to bolts not fitting bolt holes and nuts being loose.

Inspect all parts of the plow. See that the eveners and pole are in good condition. Lever dog boxes should be oiled so they will work freely. And, above all, keep the polished parts free from rust.

Opening a Land. To open a land with a riding plow, both the land wheel and front furrow wheel levers should be raised until plow opens up to depth desired. On the next round, the landside lever is adjusted to permit bottoms to cut depth desired to plow, and the furrow wheel is set level with the bottoms. Once the correct setting has been made, the plow will run level and continue cutting at uniform depth.

Many plowmen prefer to turn the first furrows in each land with a walking plow, thereby simplifying the adjustment of the riding plows.

When plowing in "lands," the bottoms should be lifted out at each end while the plow is moving. The forward motion of the plow aids in lifting.

When plowing around a field, the plow is not lifted at the corners. The driver stops a short distance from the end, turns square, and does not permit his team to start ahead until completely turned.

Hitching Riding Plows. The satisfactory performance of a riding plow depends to a great extent upon correct hitching. If both horizontal and vertical hitch adjustments are correct, the plow will run smoother, pull lighter, and do a better job of plowing than when carelessly hitched.

Fig. 34 illustrates the correct up-and-down adjustment on the vertical clevis. The correct hitch at "A" is the place where "A" is in a true line between "B" and point of hitch at the hame. When plowing deep or using tall horses, hitch at "A" should be higher than when plowing shallow or using small horses. When hitching horses strung out, the hitch at "A" must be lower than when using four horses abreast.

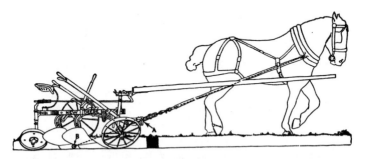

Figure 34—Illustrating the correct vertical hitch for riding plows. Straight line from center of draft, "B", to center of power at hame, passes through point of hitch at "A".

The results of improper adjustment of the vertical hitch are easily noticed.   If the hitch is too high at "A", there is a down-pull on the front end of the plow and the rear end will tend to come up.   If the hitch is too low at "A", the draft will tend to lift the front end of the plow. By changing the position of the clevis up or down one or two holes at "A", a trial will generally show which hole places the clevis in a true line of draft.

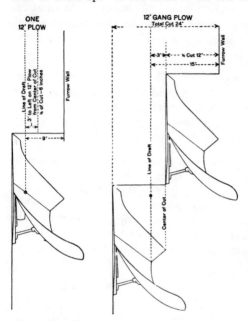

Figure 35—Finding the center of draft on any size plow is simple, if this illustration is followed.   **First**, find total cut of plow.   **Half** of total cut is center of cut.   **Measure** to left of center of cut 1/4 the width of cut of one bottom to get center of draft.

Hitches are adjustable horizontally for the purpose of accommodating the position of the horses and the various sizes and types of eveners.   Consequently, the cross hitch is very long and has a large number of hitch positions.

The operator should aim to get the horizontal hitch as near as possible in direct line between the center of draft and the center of power when the plow is running straight and horses are pulling straight ahead.

Fig. 35 shows how to determine the center of draft on any size plow.   One-fourth of cut of one bottom measured to left of center of cut of the entire plow determines center of draft. This applies to one or any number of bottoms used. Detailed explanation accompanies illustration.

"A" in Fig. 36 shows the approximate position of the evener and clevis for a four-horse strung-out hitch. To accommodate horses, the hitch is to extreme right, tending to cause side draft and "running in" of plow.

"B" shows the correct gang plow hitch for best results—the five-horse strung-out hitch. Line of draft is correct, plow runs straight and steady, and horses are not crowded in their positions.

"C" illustrates the four-horse-abreast hitch. Note necessity of hitching to extreme left to accommodate furrow horse, resulting in side draft and "running out" of plow.

Moving the hitch one or two holes to the left in "A" and to the right in "C" will lessen side draft to some extent. However, the five-horse strung-out hitch, shown in "B", should be used if conditions permit.

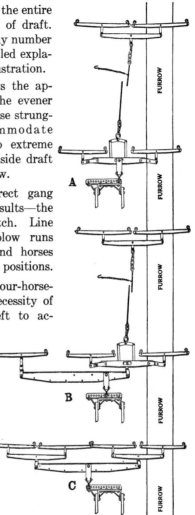

Figure 36—Showing advantages and disadvantages of three types of riding plow hitches.

Lines and Traces. Comfort to horses when plowing depends a great deal upon the adjustment of lines and traces. The lines should be adjusted so the horses cannot spread out too much, for, when the horses are spread out, they will not be well under control of the driver.

Long traces give the horses more room and tend to make the plow run steadier. Short traces do not lighten the draft and may add much discomfort to the horses.

If hip straps are used, they should be adjusted so that the loops hang free. If loops pull up on the traces, they will change the line of draft, making the plow run unsteadily and cause weight to be carried on the horses' backs.

The plow operator should keep constantly in mind the welfare of his horses as well as his plows.

Questions

1. What are the three general types of riding plows?
2. What two styles of sulky plows are in common use?
3. Describe the two-way plow.
4. Name and describe the type of gang plow in most common use in your community.
5. How is the weight of a riding plow carried when the plow is at work?
6. What tests would you make to determine whether or not the rear furrow wheel is set properly? How would you correct faulty setting?
7. What causes furrows of uneven widths and how would you correct width of furrow cut by front or rear bottom?
8. What two uses has the foot lift?
9. When would you adjust the lifting spring?
10. Describe how you would overhaul a riding plow.
11. How would you set a riding plow for opening a land?
12. What is the vertical hitch and what is its proper adjustment?
13. How do you determine center of draft on any plow?
14. What is the ideal hitch for two-bottom gangs?
15. What is the proper adjustment of traces and lines?

Tractor Plows

Types. Tractor plows, like horse-drawn plows, are built in several different styles. This is necessary to meet the requirements of tractor owners in all sections of the country.

The general farmer can buy a plow with from one to five bottoms, to suit the power of his tractor and his plowing conditions. The orchardist can get a tractor plow that will plow up close to his tree rows, under the limbs, and do an equally good job of plow-ing in general field work.

Figure 37—Single-bottom tractor plow.

Figure 38—Three-bottom tractor plow.

The two-bottom plow, illustrated in Fig. 39, is probably the most gen-erally used size and type because of the popularity and practicability of the modern general-purpose tractor, the power of which is such that the two-bottom plow makes the ideal load.

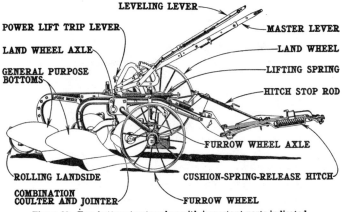

LEVELING LEVER

POWER LIFT TRIP LEVER

LAND WHEEL AXLE

GENERAL PURPOSE BOTTOMS

MASTER LEVER

LAND WHEEL

LIFTING SPRING

HITCH STOP ROD

FURROW WHEEL AXLE

ROLLING LANDSIDE

COMBINATION COULTER AND JOINTER

CUSHION-SPRING-RELEASE HITCH

FURROW WHEEL

Figure 39—Two-bottom tractor plow with important parts indicated.

Fig. 37 illustrates a single-bottom plow of similar design, built for the smaller general-purpose tractors.

In Fig. 38 and Fig. 40 are shown three- and four-bottom tractor plows. They are especially adapted to larger farms and more powerful tractors, their greater capacity being an important reason for their popularity.

Figure 40—Four-bottom tractor plow, a type coming into more general use each year in sections where large tractors are used.

The orchard tractor plow is built so that, when the plow is at work, the levers are down next to the frame of the plow, out of the way of low-hanging limbs. Its hitch is reversible and stiff, to permit a wide offset to right or left for plowing close to trees.

The two-way plow, although designed originally for use in irrigated fields where all possible moisture must be conserved, and in hilly country where it is desired to follow irregular

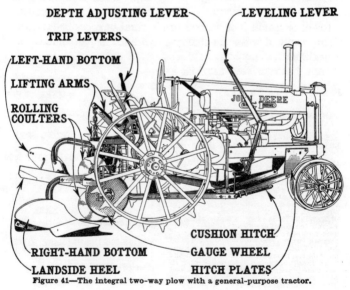

Figure 41—The integral two-way plow with a general-purpose tractor.

contours, is now used in a wide variety of plowing jobs. The market gardener, the farmer who has small, irregular fields to plow, and many others who find it an advantage to turn all furrows one way find the two-way especially valuable.

Fig. 41 illustrates a tractor two-way plow of the integral or tractor carried type. Attached to the general-purpose tractor, it becomes a compact unit with the tractor, the power lift of the tractor, operated by the tractor engine, raising and lowering the bottoms. Depth is varied, and the plow levelled by means of levers. Gauge wheels hold the plow to maximum depth set.

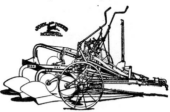

The two-way gang, shown in Fig. 42, offers the same advantages as the single-bottom and has the added value of greater capacity.

Figure 42—Tractor two-way gang plow.

In sections where unusually difficult plowing conditions prevail or where it is desired to plow at considerably greater than usual depths, the heavy-duty plow equipped with deep tillage bottoms (see Fig. 45) is available. With this type of

Figure 43—Plan for laying out and plowing a field with a tractor and plow: First—Stake out headland, about twice the length of the outfit, clear around the field. This gives all room necessary for turning at ends. Second—Plow a furrow clear around the field, as staked out. These furrows may be thrown in or out. However, it is recommended to throw in, as this leaves the headland in better shape to finish. Then, too, the head furrows mark the point to raise and lower the plow. To begin the furrow, drop the plow when the first bottom reaches the dead furrow. This leaves square headlands. Third—Open land as shown in diagram; arrows show direction of travel. After land No. 1 has become too narrow to permit the sweeping turn, swing over and open up land No. 2, plowing alternate furrows in both lands until land No. 1 is finished. Then plow No. 2 land until it is too narrow for turning; open the third land, and so on. Fourth—After all lands are plowed, start in at one corner and plow around until the entire headland is plowed and field finished.

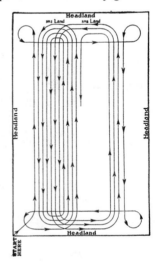

plow, good plowing as deep as 16 inches is possible and practical.

Operating Tractor Plows. The same principles that govern good plowing with horse-drawn plows apply to the

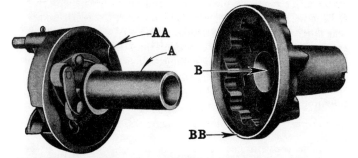

Figure 44—Details of power lift used on tractor plows shown. Wheel boxing runs on replaceable chilled-iron bearing, "A", which forms the long bearing for clutch member "AA". Wheel bearing, "B", and clutch drum, "BB", are separate from wheel, permitting replacement of these parts.

operation of tractor plows. Bottoms must be in proper adjustment, shares and rolling coulters must be sharp, wheels

Figure 45—Heavy-duty plow designed for deep work in heavy soil.

must be properly set, and the plow must be hitched correctly to insure light draft and good work.

When a new plow is delivered by the implement dealer, it is usually in proper adjustment for efficient work. If the rear wheel adjustment becomes loose and permits the landside to become lower than the rear furrow wheel, the collar on the rear furrow wheel axle should be moved up until there is one-half inch space beneath heel of the landside on the rear bottom. On plows having rolling landside in place of rear wheel, no adjustment is necessary.

The depth, or master lever on a tractor plow is used only for setting the depth. The leveling lever should be set so bottoms run level. This is necessary to do a smooth job of plowing.

The power lift, shown in detail in Fig. 44, is built into the land wheel of the plows shown. It is simple and positive. The operator simply pulls on the trip rope and the bottoms are lifted high and level. When the turn is completed, the bottoms are lowered by pulling the rope in the same fashion.

The lifting spring should have proper tension. If the spring has too much tension, it will cause the land wheel to slide when plow enters the ground. If not tight enough, the land wheel may slide a little when plow is being lifted. Adjust by loosening or tightening.

The hitch for tractor plows should be one of two types—the pin-break type, equipped with medium strength pin, or the cushion-spring-release type. In the first type, the wood pin breaks when a field obstruction is met; with the cushion-

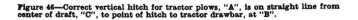

Figure 46—Correct vertical hitch for tractor plows, "A", is on straight line from center of draft, "C", to point of hitch to tractor drawbar, at "B".

spring-release type, the load is carried on a heavy coil spring which is compressed when the load becomes too great, thereby permitting a latch to withdraw, releasing the plow.

How to Plow a Field.   A definite plan of plowing a field is necessary to good plowing and time-saving with a tractor plow.   The plan of opening and plowing in lands, as shown in Fig. 43, has been found very pratical.   It reduces to a

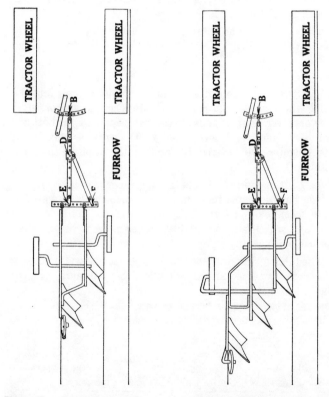

Figure 47 — Correct horizontal hitch for two-bottom tractor plows.

Figure 48—Illustrating how to adjust horizontal hitch for three-bottom tractor plows.

minimum the amount of time spent in turning and moving with the bottoms out of the ground. Details of this plan are given with the illustration.

Care of Tractor Plows. What has been stated in previous paragraphs on the care of riding plows and plow bottoms applies to tractor plows as well. Liberal use of oil and grease on working parts, and on bottoms when not in use, adds years to the life of the tractor plow.

It is a good plan to go over the plow before storing and to put it in condition for the next season's work. Sharpen shares or get new ones if the old ones are too badly worn; tighten bolts, replacing where needed; readjust or replace any parts that show excessive wear. Reconditioning plows during slack seasons saves time in the busy plowing season.

Hitching Tractor Plows. The most important factor in tractor-plow operation is correct hitching. Draft, quality of work, and ease of operation depend to a great extent upon the hitch.

In hitching the tractor plow, as in the hitching of any wheel plow, keep in mind the fact that when the plow is at work, the

Figure 49—Good plowing is the aim of every good farmer.

wheels serve to carry the weight of the plow and the earth
being turned, and to act as gauge wheels to keep the plow
bottoms level at uniform depth.   The principles of center of
load in relation to center of power must be observed just as
in the case of the walking plow.   Figs. 46, 47, and 48 illustrate the correct horizontal and vertical hitches for two- and
three-bottom tractor plows.

Fig. 47 illustrates the correct horizontal hitch for two-
bottom tractor plows.   Attach drawbars at "E" and "F", as
shown.   Adjust drawbar at "D" to make front bottom cut
right size furrow slice.

Fig. 48 shows proper horizontal hitch adjustment for three-
bottom tractor plows.   With drawbars attached as shown at
"E" and "F", width of cut of front bottom is regulated by
adjustment at point "D".

Fig. 46 shows the correct vertical hitch of tractor plows.
The line from center of draft, point "C", should pass through
hitch at "A", to point of attachment to tractor.   If this line is
broken by raising hitch at point "A", the resulting down-pull
on front of plow tends to raise rear of plow out of the ground.
Likewise, too low a hitch at "A" will raise front end of plow.

Careful adjustment of the hitch insures an even-running
plow, a good job of plowing.   (See Fig. 49.)

Questions

1. *Describe several types of tractor plows and tell the conditions to which
    they are adapted. What type is most generally used in your community?*

2. *How would you adjust the rear wheel of a tractor plow for best work?*

3. *What is the purpose of the depth lever?   The leveling lever?*

4. *What is the purpose of a cushion-spring-release hitch?*

5. *How would you open a field for plowing with a tractor?*

6. *How would you obtain the correct vertical hitch on a tractor plow?
    The correct horizontal hitch?*

Disk Plows

Types and Uses. Disk plows are used in territories where soil conditions are such that moldboard plows will not operate to best advantage. They work well in soil so dry and hard that moldboard plows cannot penetrate, and in sticky soils, such as waxy muck, gumbo, and hardpan, where moldboard plows will not scour well. They are also used to advantage in very loose ground and in stony and rooty land. Fig. 52 illustrates the way a disk plow cuts and turns the furrow.

Disk plows are built in both tractor-drawn and horse-drawn types in styles to meet varying requirements. The plow shown in Fig. 53 is a heavy-duty type, built especially to meet hard soil conditions where penetration is the big problem and where it is desired to plow extremely deep.

The disk plow illustrated in Fig. 50 is a popular style for use with tractors. In addition to this style which is built in several sizes, there are smaller plows of similar design (see Fig. 52) built for use with the modern general purpose

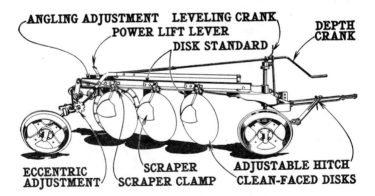

Figure 50—Tractor-drawn disk plow, important parts named.

type tractor. For use with larger, heavier, more powerful tractors there are several types of heavy-duty and extra heavy-duty plows (Fig. 53) designed to meet the toughest plowing jobs encountered.

Operation and Adjustment. Since all disk plows operate on practically the same principles, the tractor-drawn plow, shown in Fig. 50, will serve as a basis for this discussion.

Disks must operate at the same level to do an even job of plowing. The rear wheel may be raised or lowered by the adjustment of a simple eccentric to fix its proper relation to the level at which the disks run.

The front furrow wheel can be set to run straight or at a slight angle, according to conditions, by adjusting the hitch bars. When opening lands, it is sometimes necessary to raise the furrow wheel higher than ordinarily required, by turning the screw crank depth adjustment.

The land wheel, which includes the power lift in the tractor plows, should run straight.

Scrapers must be set so they barely touch near the center of the disk, with the wing one-fourth inch from the surface. If set too close, excessive friction is created; if too far, they will not do a good job of cleaning.

Adjusting Load to Power. Practically all disk plows are convertible in number of disks used and the cut per disk,

Figure 51—Showing cross-sectional view of roller bearing disk bearing with disk removed. Tapered roller bearings installed at spindle end and shoulder carry the pressure of plowing with a minimum of draft. Expanding felt collar and overlapping flange on the disk spindle make this bearing dirt-proof.

making it possible for the operator to fit the total width of cut
to the power available and to soil conditions.   In some of the
larger heavy-duty plows, one disk may be removed and the
remaining disks respaced in which setting the plow cuts same
width as before, but each disk takes a wider cut.

The standards to which the disks attach on the plow
shown in Fig. 50 are bolted between two frame bars.
Width of cut per disk is increased or decreased by changing
the angle of the frame.   This is done by loosening four bolts
which attach the rear frame casting to the angle frame bars,
and shifting the rear frame.

Care Important.   The disks of a disk plow are important
factors in its operation.   If they are kept sharp, polished,
and properly adjusted, they cut and turn the furrow slice
with the least possible draft.   If neglected, they soon become
the source of penetration troubles and poor work—the disk

Figure 52—Disk plow at work, showing how the disks cut and turn the soil.

surfaces should be oiled when not in use.   The wheel boxings should be kept well greased with a good grade of hard oil.

The roller bearing disk bearing, cross-section of which is shown in Fig. 51, does not require adjusting; it is lubricated by means of a pressure lubrication fitting.

Hitch Adjustments.   The same relation between power and load that governs the hitching of moldboard plows applies to disk plows.   The vertical hitch must be such that the pull of tractor or horses holds the plow steadily to the depth set without raising or lowering the front end.   A trial hitch and observation of results will indicate the proper hole in which to hitch.

The horizontal hitch should be in direct line with the center of draft which is approximately the center of the total width of cut on any type of disk plow.   Adjustment one or two holes to right or left may be necessary to accommodate horses or tractor, in which case the furrow wheel may require adjustment so that the front disk cuts proper width.

Figure 53—Preparing a seed bed, in tough conditions, with a heavy-duty disk plow capable of working as deep as twenty inches.

Questions

1. *Under what conditions are disk plows used?*
2. *Name and describe two types of disk plows.*
3. *What is necessary to do an even job of plowing?*
4. *How would you increase or decrease the width of cut per disk?*
5. *What are the most important points in caring for a disk plow?*
6. *How would you obtain the proper hitch adjustment?*

Wheatland Implements

The necessity for implements of big working capacity to aid in cutting production costs in winter-wheat districts has brought out implements of special design. The listing plow,

ridge burster, and the disk tiller have doubled and trebled the working capacity of farmers who formerly depended upon regular plows for preparing their wheatland. These implements not only have aided in the production of wheat at lower costs,

Figure 54—Listing plow for use with tractor.

but in some of the more sandy and dry sections have established new cultural methods that save moisture and increase yields.

New to the regions of limited rainfall is the damming lister, designed to place fields in ideal condition to hold and store moisture and prevent soil erosion by wind and surface water run-off. Fig. 58, showing a damming lister

Figure 55—Ridge burster in operation, showing how it levels down the ridges made by the listing plow.

at work, shows clearly the topographical condition of fields worked with this equipment.

Listing Plow and Ridge Burster. These two implements are used in conjunction—the listing plow (Fig. 54) throws up ridges, forming a surface mulch and moisture trap, and the ridge burster levels the field. Listing is usually done immediately after harvest, and ridge bursting just before planting. Fig. 55 shows the ridge burster bursting ridges previously made by a listing plow.

The principles of operation, care, and adjustment of all types of listing plows are so similar to those of other moldboard plows that the previous text matter on this subject may be considered sufficient.

The ridge burster is comparatively simple to operate and adjust. The disks may be moved in or out on the crossbars or angled to their proper positions for best results by making simple clamp adjustments. Levers provide adjustment for desired depth. Thorough and regular oiling of the disk bearings will add years to their life. Disks should be protected from rust between working seasons by a covering of oil.

The Disk Tiller. The disk tiller (Figs. 56 and 57) is similar in operation to the disk plow. It is widely popular among winter-wheat growers and is gaining popularity with general farmers in practically all parts of the country. Once over

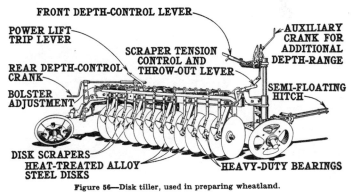

Figure 56—Disk tiller, used in preparing wheatland.

after harvest puts the land in condition for planting in some sections; in other sections, it is common practice to till the ground once after harvesting and again just before planting.

Although designed primarily for use in the large wheat-growing sections, the disk tiller has proved so successful in many general farming sections that its use is no longer restricted to one territory—a fact which has resulted in the development of tillers in various sizes and in modified types to take care of a wide range of uses in weeding, seed bed preparation, working fallow ground, etc. The kind of work a disk tiller will do depends much upon the condition of its disks and bearings. Disks must be sharp to cut the stubble and trash, and their bearings well oiled to withstand the heavy work to which they are subjected.

A long lever at the front, supplemented by an auxiliary crank for additional range, controls depth of the front furrow wheel. This control device provides plenty of adjustment for opening up land. An easily operated screw crank at the rear acts upon both the land and rear wheels. Once leveled to the proper depth, no further adjustment is necessary as the power lift raises the disks or drops them to work when turning at the headlands. The correct hitch is on a straight line between center of draft and point of attach-

Figure 57—Disk tiller at work. Note how stubble is mixed with the surface soil, forming a mulch.

ment to tractor. A trial will usually show where the hitch should be.

A bolster adjustment on the rear of the tiller shown on the preceding pages provides adjustment for changing the angle at which the disks work. By means of this adjustment, any working angle required to meet conditions ranging from soft, loose soil, to hard, dry ground can be obtained. The rear frame is definitely marked with settings for three soil conditions: "soft or stony ground," "medium ground," and "hard ground."

The rear gang of three disks on the tiller, shown in Fig. 56, may be removed to adapt the tiller to small-powered tractors or where field conditions are unusually difficult. This reduces the width of cut, lessening the power required.

Questions

1. *Describe a listing plow and tell of its uses.*
2. *What is the purpose of the ridge burster and how is it adjusted?*
3. *Describe the disk tiller, its advantages and uses.*
4. *How is depth controlled on the tiller shown?*
5. *What is the purpose of the bolster adjustment?*
6. *What is the purpose of the damming lister and where is it used?*

Figure 58—Damming lister at work on the Great Plains.

Chapter II.
DISK HARROWS

The function of the disk harrow is to pulverize and pack the soil, leaving a surface mulch and a compact subsurface. It is used to good advantage before plowing to break the surface and mix the trash with the topsoil, and after plowing to pulverize lumps and close air spaces in the turned furrows. Although it is not to be found on every farm, there are few general farmers who would not profit by its use. Fig. 59 illustrates the value of a disk harrow when used before and after plowing.

Sun-Baked Stubble Land

Plowed, but Not Disked

Disked After Plowed, but Not Before. Notice Air Spaces

Disked and Then Plowed. Good Contact with Subsoil

Disked Before and After Plowed. The Ideal Seed Bed

Figure 59—Drawings illustrating the value of disking both before and after plowing.

Types of Disk Harrows. Disk harrows are made in single-action and double-action types. Most of the double-action harrows and some of the single-action harrows are designed for use with tractor only. Others may be used with either horses or tractor, by changing the hitch. Some types of single-action harrows can be converted into double-action harrows for use with horses or tractor by adding a rear section.

Requirements for Good Work. To do a good job of disking, the disk harrow, first of all, must penetrate well and evenly over its entire width. In the case of two-section machines, both sections must meet these requirements, the disks of the rear section cutting the ridges left by the front disks instead of trailing in their furrows.

Flexibility has much to do with even penetration and good work. When the gangs of each section work independently, one gang may pass over stones or stumps and conform to irregularities in the surface of the field without hindering the work of the other gangs.

Operation and Adjustments. Penetration of a disk harrow is obtained by angling the disks, the angle necessary for good work depending upon the condition (or texture) of the soil and the amount of trash to be cut. On most disk harrows, provision is made for angling the disks for maximum penetration which is obtained at an angle of approximately 20 degrees.

A recent development in tractor harrows is the tractor control, whereby the gangs are angled or straightened entirely by the forward or backward travel of the tractor. Angling the gangs of both the single- and double-action harrows (Figs. 60, 61, 62, and 63) is accomplished by means of trip ropes

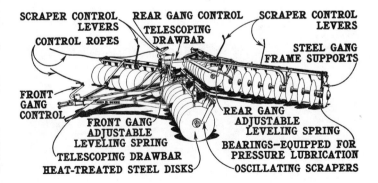

Figure 60—A double-action tractor-controlled disk harrow.

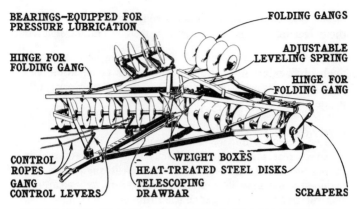

BEARINGS–EQUIPPED FOR
PRESSURE LUBRICATION

FOLDING GANGS

HINGE FOR
FOLDING GANG

ADJUSTABLE
LEVELING SPRING

HINGE FOR
FOLDING GANG

CONTROL
ROPES
GANG
CONTROL LEVERS

WEIGHT BOXES
HEAT-TREATED STEEL DISKS
TELESCOPING
DRAWBAR

SCRAPERS

Figure 61—Single-action tractor-controlled harrow with end gangs folded over.

within reach of the tractor operator. Both gangs of the single-action harrow, and the front gangs of the double-action harrow, can be straightened on the forward pull if in danger of stalling the tractor; to angle these gangs, the latch is released and the tractor backed against the stiff hitch until the desired angle is attained. To angle the rear gang of the

Figure 62—A single-action tractor-controlled harrow working a strip twenty-one feet wide.

double-action harrow, the latch is released while the harrow moves forward.

Front and rear gangs of the double-action harrow can be angled to the proper degree or straightened independently of each other.

The end gangs of the single-action harrow, shown in Fig. 61, can be folded over when going through gateways, or to provide additional weight for better penetration in difficult conditions. In this way, a 15-foot harrow can be narrowed to 10-1/2 feet, and a 21-foot harrow to 14 feet. The single-action harrow may be used for double-disking by lapping half the width of the machine each time across the field.

The scrapers are adjustable to suit soil conditions. Scrapers on both sections of the double-action tractor harrows shown can be oscillated by means of ropes, without leaving the tractor seat. Pressure on foot levers on the horse-drawn models oscillates scrapers.

Figure 63—A tractor-controlled double-action disk harrow doing good work.

Provision is made for locking scrapers at the edges of disks, or for locking them away from disks when not needed.

Good Care Lengthens Life. The efficiency and length of service of a disk harrow depend upon the care given it.

First in importance is thorough greasing of the bearings. Many of the modern disk harrows, especially the tractor types, are equipped with fittings for pressure lubrication, making it an easy matter to keep the bearings well oiled. Where hard oilers are used, cups should be kept full of a good grade of hard oil and should be turned down at regular intervals. The bearing bushings of hard maple are oil-soaked before they are assembled, giving them long life. They are easily replaced when worn.

A good cutting edge on all disks is desirable, especially in hard ground and trashy conditions. Most disk blades are now made of tough steel, then heat-treated to hold a long-wearing edge.

During slack seasons, go over the entire disk harrow, tightening bolts, replacing worn parts, and getting the implement ready for the next season's work. Keep disks well greased with a good hard oil when harrow is not in use.

Questions

1. *What is the function of the disk harrow?*
2. *Why should it be used both before and after plowing?*
3. *What is the first requirement of a disk harrow?*
4. *What is meant by "flexibility" and why is it desirable?*
5. *How is a disk harrow made to penetrate?*
6. *Describe the manner of angling and straightening the gangs on the tractor-controlled harrows.*
7. *How are the disk scrapers oscillated?*
8. *Why is it necessary to keep the disks sharp?*

Chapter III.

HARROWS, PULVERIZERS, AND FIELD CULTIVATORS

Methods of finishing the seed bed vary according to soil conditions and established practices in different sections of the country. While the spike-tooth harrow is used in practically every section, the spring-tooth harrow, pulverizer, and field cultivator are not in such general use. For that reason, the discussion of those implements will be brief.

Spike-Tooth Harrows. Fig. 64 shows a popular style of spike-tooth harrow. The operation and adjustment of a harrow of this type are comparatively simple, there being no field adjustment other than setting the slant of the teeth with the levers provided. This adjustment is governed entirely by field conditions.

Each tooth of this harrow is held between the two notched, semi-oval frame bars by the heavy bolt which creates a tension, thereby locking the tooth to position and preventing the nut from coming loose. When one side of the tooth becomes worn, the nut may be loosened and the tooth turned to present a

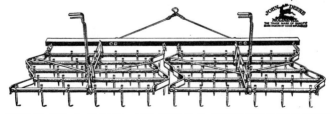

Figure 64—Spike-tooth harrow, with detail above showing how tooth is locked between bars.

new cutting edge. Teeth may also be removed for sharpening.

Spring-Tooth Harrows. The fact that spring-tooth harrows will penetrate to a greater depth than spike-tooth harrows makes them better

Figure 65—A three-section, tractor-controlled spring-tooth harrow.

adapted to the requirements of certain sections. They are also used with great efficiency in the eradication of obnoxious weeds and grasses.

Fig. 65 illustrates a tractor spring-tooth harrow that is controlled from the tractor seat. Trip ropes are provided for dropping the teeth to work and raising them out of the ground, depth being determined by previous setting. In

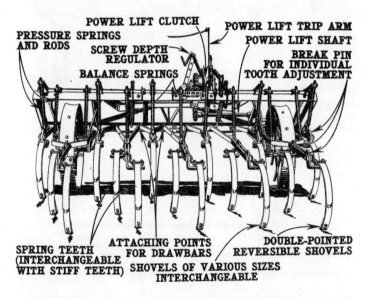

Figure 66—Field and orchard cultivator with tractor hitch and screw-type depth regulator.

addition to this depth adjustment, the individual teeth may be adjusted in the clamps for finer depth adjustment or to vary the penetration.

Several styles of teeth may be obtained for different purposes such as cultivating alfalfa and digging out quack grass roots, in addition to the type used as a seed bed finisher.

Field and Orchard Cultivators. Because of its great diversity of uses, the field cultivator is used extensively in both the United States and Canada. The type shown in Fig. 66, low-down, with wheels set inside the frame, is ideal for working close to trees, fences, and ditches, in both field and orchard.

It is used for preparing fall-plowed land for spring seeding and for tillage work in regions where summer-fallowing is practiced. It does good work as a weed destroyer, eradicating quack grass, thistle, wild morning-glory, and other weeds; it is also an efficient alfalfa cultivator.

Figure 67—Soil pulverizer—double-roller, horse-drawn style.

The field cultivator shown in Fig. 66 may be used with either stiff or spring teeth and with different kinds and sizes of shovel points. Either tractor or horse hitch may be obtained with the field cultivator.

The field cultivator is simple to operate, there being few adjustments. Teeth are raised and lowered by means of a power lift operated by a trip rope from tractor seat. Crank-type depth regulator gives tractor operator constant depth control. Hand levers serve the same purpose on horse-drawn, hand-lift machines.

Frequent sharpening of the shovels, and keeping them coated with a good hard oil when not in use, will aid in making the cultivator pull light and do good work.

Soil Pulverizer. Ideal for finishing the seed bed and valuable for use in growing crops, the soil pulverizer, or packer (Fig. 67) is more widely used each year. It is simple and easy to operate. There are no adjustments and few parts that ever need replacing, with exception, possibly, of the oil-soaked wood boxings which are easily removed and replaced at small cost.

Questions

1. *What implements are used for finishing seed beds in your community?*

2. *What adjustment can be made on the teeth of the spike-tooth harrow shown?*

3. *To what conditions are spring-tooth harrows especially adapted?*

4. *Describe a field cultivator and its uses.*

5. *Why is the soil pulverizer a valuable implement for finishing the seed bed?*

Part Two

PLANTING

The necessity of planting all crops at the proper depth and uniformly distributing the right amount of seed to suit soil conditions is quite apparent to all who have had experience in growing products of the soil.

If seed is planted too deep or too shallow, too thick or too thin, if the row planter skips hills or the grain drill leaves strips unplanted, the yield is bound to be reduced accordingly. If planted accurately, with the per acre quantity carefully measured to suit the richness of the soil, maximum yields will result, provided, of course that other conditions and practices are correct.

The farmer who understands and gives careful thought to the adjustment and operation of his planting equipment will profit greatly. The operation of planting equipment should be carefully studied by the students of agriculture and farm mechanics.

Chapter IV.

GRAIN DRILLS

Grain drills have been greatly improved during the past few years. Perhaps the most notable improvement is the rust-resisting galvanized steel box. Not only does the steel construction give greater strength and durability, but it also makes possible an enormous increase in capacity; in some cases as much as 72% more than that of the wood box.

Like plows, grain drills are built in many different styles with a variety of equipment to meet conditions in every section of the country. In some sections, the single-disk furrow opener will work better than the double-disk, while other conditions may demand a hoe-type opener. The big farm regions require big tractor-drawn drills (Fig. 75) while the

small farmers of the East or South need only the smaller
sizes, many using the one-horse drill (Fig. 68). Many farmers
in the larger wheat-growing sections prefer the double-run
feed type of drill. In semi-arid regions, where every available
bit of moisture must be conserved, the semi-deep furrow drill
with its large disks solves the problem by placing the seed
considerably deeper than the ordinary drill, thereby assuring

contact with the moist soil found at
greater depth. In territories where
soil blowing is a serious problem, the
deep furrow drill with moldboards
which throw the soil one way (Fig.
74) serves the purpose of deep planting and, at the same time, leaves the
surface soil ridged to prevent or reduce soil drifting and seed blowing.

Figure 68—One-horse drill,
popular with farmers who
have small acreages.

The lister drill, shown in Fig. 72, is still another variation of
the grain drill. Its value lies in the fact that it makes deep,
wide furrows and high, rounded ridges thereby preventing or
reducing soil blowing. But whatever the preference may be,
or conditions demand, it is usually found that manufacturers
build a drill that meets the requirements satisfactorily.

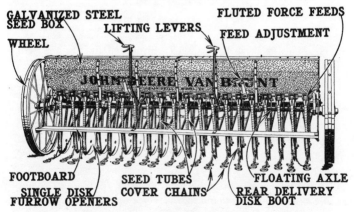

Figure 69—Rear view of grain drill, equipped with single-disk openers.

Types of Drills. There are three principal types of end-wheel grain drills—the fluted feed, the double-run feed, and the combination grain and fertilizer drill. Of these, the fluted feed (Fig. 69) is in most general use, although in some sections, the other styles are used almost exclusively. The Pacific-Northwest states favor the double-run feed drills, while eastern and southern farmers find the combination fertilizer and grain drill most practical for their conditions. Then, there is the low-down press drill having press wheels that firm the soil over the seed to prevent it from blowing. The plow press drill, similar in construction, is attached behind the plow. A pulverizer may be used between the plow

One-piece seed cup
Latch in upper left notch

Latch in right-hand notch

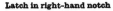

Position 1 — All gates up with latches in top notch at left side to sow all grains, small seeds, kafir corn, and beets.

Position 2—Fasten all latches at right side to sow peas, common beans, soy beans, corn, and extra large quantity of trashy oats.

Latch in lower left-hand notch   Feed roll

Feed shaft
Feed cut-off

Special soy bean gate in position.
Regular gate down

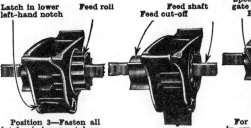

Position 3—Fasten all latches in lower notch on left side to sow soy beans, marrow fat, or kidney beans.

Position 4—Drop gates to clean feeds.

For soy beans and rice in small quantities, insert special gate with cotter. The feed gate must be down.

Figure 70—Detail of the fluted feed showing different settings that can be made for planting seeds of various sizes. Quantity is controlled by shifting the feed roll and feed cut-off to permit more or less of the feed roll to turn within the feed cup. This is done with the feed shaft shifter.

and the drill, making a three-way hook-up for plowing, pulverizing, and seeding, all in one operation.

Fluted Feed. The fluted force-feed consists mainly of a feed roll, feed cut-off, seed cup, and an adjustable gate. The feed roll turns with the shaft, forcing the grain out over the feed gate which is adjustable for different sizes of seeds. The feed cut-off and the feed roll shift with the feed shaft, and their position determines the quantity of seed sown. The one-piece seed cups aid in maintaining accuracy because they do not become loose and get out of line.   (See Fig. 70 for detailed explanation.)

Setting Fluted-Feed Drills for Quantity. The first adjustment in using any drill is to set it to sow the desired quantity per acre.   This is done on fluted feed drills by adjusting the feed shaft and the gates on the feeds to suit the size of seed and the quantity to be sown.

The setting of the adjustable gate force-feeds according to size of seed is described under the illustrations in Fig. 70. Before putting grain in the box, all gates should be let down as in No. 4, Fig. 70, and all grain and accumulations

Figure 71—Large-sized tractor-drawn drill used in big farm sections.

cleaned out.   The latches on all feeds must be kept in same position while seeding to insure uniform planting.

The feed adjustment, or feed shaft shifter, moves the feed rolls and feed cut-offs to permit more or less grain to be forced out by the feed rolls.   There are two of these shifters on drills having more than eight disks, one for each half of the drill.   Both must be kept in the same position on the seed index plate which is provided with a row of notches to hold shifters in position.   These notches are numbered by the figures which are immediately above them.   Figures above at left of notches indicate the amount of flax and alfalfa—in pounds—to be seeded per acre.   Figures below notches indicate amount of oats, barley, wheat, and peas—in pounds—to be seeded per acre.

Double-Run Feed Drills. The double-run feed drill gets its name from its type of feed, illus-

Figure 72—The lister drill with deep furrow openers and press wheels.

tration of which is shown in Fig. 73. The feed and the mechanism which drives it constitute the principal differences

between this type of drill and the fluted-feed drill, shown in Fig. 69.

Fig. 73 shows two views of the double-run feed. It consists mainly of a feed wheel and a feed gate. The wheel is smaller on one side, for use in planting small seed. The large side is used for planting oats, barley, treated wheat, peas, beans, and other large seeds.

The adjustable gates, which are inside the seed cups, regulate the side of the feed openings, there being five different positions at which they can be set— three on the large side and two on the small side. These five gate adjustments provide five different quantity adjustments for each one of the five multiple gears, making a total of 25 different quantities in which seed may

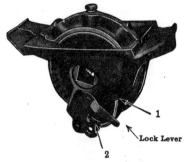

Arrows point to lock lever on small side of feed, used for regulating inside quantity seed gate, and to the positions at which it may be set. Numerals indicate the two positions, No. 1 being for large quantities, and No. 2 for small quantities.

Showing adjustable gate inside the seed cup for regulating size of the feed opening to handle different quantities of seed. Gate is set on position 2.

Figure 73—Detail view of double-run feed, showing large and small sides.

be planted without changing gears.　By reversing the intermediate gear, 25 additional quantity adjustments are provided—50 in all.

Fertilizer-Grain Drills. This type of drill completes four operations at once.　It pulverizes the soil, plants seed, distributes fertilizer, and covers both.　Farmers who find it necessary to sow fertilizer on their fields are able to make a big saving in time and money by using a fertilizer drill, sowing the fertilizer and seed in one operation.

Fertilizer drills have two distributing units—one for seed and one for fertilizer—although both seed and fertilizer are released through the same tube.　The planting unit, which

Figure 74 — Furrow opener with moldboard and seed deflector as used on deep furrow drill.

consists of the regular fluted feed drill mechanism illustrated in Fig. 70, is built into the front half of the seed box.　The fertilizer feed distributes any amount of fertilizer from 24 to 1225 pounds per acre.　Star feeder wheels rotate in the fertilizer box and cause an even flow of material into the seed tubes.

Because of the increasing use of highly concentrated fertilizer, a number of drill manufacturers are now offering a special fertilizer attachment for all types of drills which keeps the fertilizer from coming into contact with the seed. The fertilizer is released through separate tubes and deposited in the rows, but there is always a layer of soil between the fertilizer and the seed.　The depth is controlled by a simple adjustment on the fertilizer boot.

Operation and adjustment of fertilizer-grain drills, with exception of the fertilizer unit, is the same as given for fluted feed drills in a preceding paragraph.

Calibrating Grain Drills. The operator should be sure to have his drill set properly before starting to sow. If there is doubt in his mind as to the accuracy of his machine, he may make the calibration test which follows:

To check the accuracy of a grain drill, jack it up in working position, fill the box with grain, place a canvas in position to catch the grain, and set the gates and feed shifters properly. Find the total width of strip planted each time across the field. Divide 43,560—the number of square feet in an acre—by the width of strip planted and you have the length of a strip necessary to make one acre. Then find the number of times the drill wheel must turn in going this distance by dividing the number of feet by the circumference of the wheel.

Tie a cloth to a spoke of the wheel and count the revolutions as you turn the wheel, turning at about the same speed it would travel at work. You need not sow a whole acre—one-fourth of an acre is sufficient for the test.

When the correct number of revolutions has been made, weigh or measure the grain on the canvas and check it with the adjustment on the feed-shifter scale. If the drill is planting more or less than it should, the difference can be taken care of by adjusting the feed shifters.

Figure 75—Drilling with tractor power on a large Western ranch.

Field Operation. To do a good job of sowing, the drill must be run steadily and evenly. Swinging of poles or unsteady driving causes bunching of seed and results in reduction of yields.

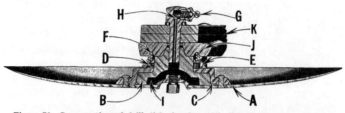

Figure 76—Cross section of drill disk showing "A", disk blade; "B", disk bearing; "C", bearing case; "D", felt washer; "E", hard iron dust cap; "F", dust cap spring; "G", Alemite fitting; "H", oil passage; "I", oil reservoir; "J", disk boot casting; "K", drawbar.

The depth of seeding over full width of the drill is controlled by the lifting levers and by a pressure spring on each furrow opener. When pressure is applied to the furrow openers, it should be uniform. Uniform pressure can be gained only by having both lifting levers in the same notch and having the pressure on all springs the same. The pressure on each furrow opener is adjusted by raising or lowering the adjusting collar on the pressure rod.

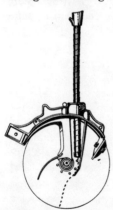

The tilting levers on rear of poles provide easy adjustment for proper relation between penetration and depth of planting when using any type of furrow opener.

Disk scrapers should be adjusted as lightly as practical and disengaged entirely, when possible, to prevent wear.

Figure 77—Cutaway view of double-disk opener showing how seed is protected between disks until it reaches the open furrow.

The land measurer is provided to measure the number of acres covered by the drill. On some drills, the land measurer is driven from the main axle; on others, from the feed shaft. To set it when starting a new field, press top of measurer in to force the bottom gear out of contact with worm gear on feed shaft or axle. Turn bottom gear to right —about 1/8 of an acre— to disengage fraction gear from acre gear. Move the indicator to largest number on acre dial and turn bottom gear to left, with indicator on fraction dial in upward position.

Figure 78—Shoe type of furrow opener.

Care of Drills. The drill should be cleaned and put in condition for the next season's seeding be-

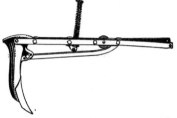

Figure 79—Hoe type of furrow opener with spring-trip.

fore it is stored. All seed should be cleaned out, the disks or other opener surfaces cleaned and oiled, and the machine put under shelter. Good treatment prolongs the life of the drill.

Most drills are equipped with fittings for pressure gun lubrication. The disk bearings should be kept oiled thoroughly with oil or grease of proper viscosity as listed in the manufacturer's instruction book. Bear in mind that the disk bearings operate largely below the surface of the ground, and for that reason it is highly important to keep the oil chamber well filled with oil of proper grade. See the cross section of disk and bearing, Fig. 76. Double-disk openers are oiled from the top of the boot.

Types of Openers. Fig. 77 shows a cutaway view of a double-disk furrow opener, illustrating how seed is protected in seed tube and between disks until it reaches the bottom of the furrow.

The shoe type of furrow opener, shown in Fig. 78, works well in loose soils where the greater penetration of a disk opener is not required.

Fig. 79 illustrates the hoe-type of opener which is especially adapted to seeding rocky soils.

Fig. 80 shows a single-disk opener with pressure spring, scraper, and disk boot.

The single disk deep furrow opener is used with 12-, 14-, or 16-inch spacing to make wide, deep trenches and ridge the soil to catch the moisture and prevent the soil from blowing. It is used most widely in winter-wheat sections.

Figure 80—Single-Disk Opener.

The deep furrow opener with moldboard and seed deflector is shown in Fig. 74.

All of these types of furrow openers are interchangeable.

Questions

1. *What is the relation between planting and crop yields?*
2. *What is considered good planting of the crops grown in your community?*
3. *Describe the fluted feed and tell how you would adjust it for quantity.*
4. *Describe the double-run type of feed and its adjustment.*
5. *What are the advantages of a fertilizer drill?*
6. *How and why would you calibrate a grain drill?*
7. *How is depth of sowing controlled?*
8. *What are the important points to remember in caring for grain drills when in operation and between seasons?*
9. *What type of furrow opener is used in your community and why?*

Chapter V.

CORN PLANTERS

Accurate planting has more to do with the yield of corn and other row crops than any other mechanical factor. If hills are crowded, barren stalks and small ears result. If hills are missed, or if less than the desired number of seeds are dropped, time and land are wasted.

The illustrations in Fig. 81 picture the results of accurate and inaccurate planting when three stalks per hill is ideal for soil conditions. In poor soils, two stalks per hill is sufficient, while in very rich loam soils four stalks will do

A, three kernels in each hill produce a perfect stand and yield three good ears.

Figure 81—Illustrations A, B, and C contrast the usual results of accurate and inaccurate planting when the soil will support three stalks of corn and produce three good ears in each hill.

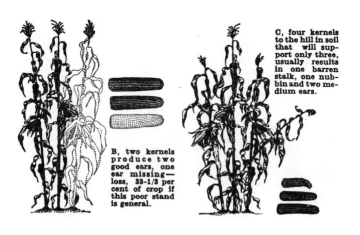

C, four kernels to the hill in soil that will support only three, usually results in one barren stalk, one nubbin and two medium ears.

B, two kernels produce two good ears, one ear missing—loss, 33-1/3 per cent of crop if this poor stand is general.

well. The inaccurate spacing of drilled corn will result in the same losses as pictured in Fig. 81.

Drop and Seed Plates. The accuracy of a corn planter depends upon the accuracy of the drop and the selection of seed plates best suited to the size of seed to be planted—taking for granted, of course, that seed is of uniform size and that dirt has not clogged the seed passages.

There are two types of corn drops—the accumulative, and the full hill. The accumulative drop is generally conceded to be more accurate because it takes one seed to each cell in the seed plate and then counts out the number of seeds to a hill as desired. A full hill drop planter takes all the seeds that make up the hill into

Figure 82—Top view of corn hopper bottom showing seed plate in position.

one cell. It is claimed, and probably rightly so, that it is easier to get one seed in a cell each time than it is to get more than one, the same number each time. The accumulative drop is described in the following paragraphs.

Fig. 83 shows a cross section of a seed hopper bottom, showing seed plate in position and sloping surface of the bottom. The weight of the seed causes it to move to the sides and enter the openings in seed plate. Fig. 82 shows top view of the hopper bottom.

The assembly of the hopper, seed plate, and bottom false plate is shown in F i g. 8 4. This also illustrates how seed plates are removed by tipping the hopper forward

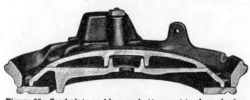

Figure 83—Seed plate and hopper bottom cut to show sloping hopper bottom, sloping hopper wall, and oblique seed plate.

and releasing the spring latch that holds the bottom plate in place, without removing seed from hopper. Extra-wide seed is accommodated by reversing the false bottom plate as indicated in the drawing.

Seed plates are now available for seed of any size from kafir to lima beans, including a full range of plates for handl-
ing the various hy-
brid strains of corn.
Fig. 85 illustrates the importance of selecting the right seed plates for corn by fitting the seed to be planted in the seed cells, as shown. If cells are too large, two ker-

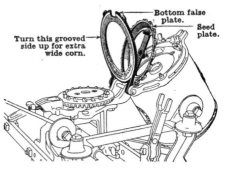

Turn this grooved side up for extra wide corn.

Bottom false plate.

Seed plate.

Figure 84—Assembly of the hopper, seed plate, and false plate showing how seed plates are removed by tipping hopper forward.

nels may pass into one cell, resulting in overcrowding the hill; if too small, less than the wanted number of kernels will be dropped.

Checking or Drilling. The horse-drawn corn planter, shown in Fig. 86, and the four-row tractor-drawn planter, shown in Fig. 89, can be used for either checking or drilling. Adjustment for planting two, three, or four kernels to the hill when checking, is made by moving the variable drop shift into the proper notch with the foot.

To change from checking to drilling, hook the foot drop rod over the small cast-
ing on the rocker shaft, using the notch nearest the end, then press down on the foot drop lever until it locks automatical-

A   B   C

Figure 85—Illustrating how to select proper seed plates for an accumulative drop. Corn must fit the cells of the plate like the kernel marked "B". If kernels are too large as "C",or too small as "A", a plate having cells that will hold one kernel at a time should be selected.

ly in the position shown by the shaded portion in Fig. 87. The foot drop lever is not used in checking.

Seeds can be planted from 4-1/2 inches to 258 inches apart when drilling. This wide range can be obtained by using plates having from two to twenty-four cells, setting the variable drop shifting lever on two, three, or four, or using the drive chain on the large, medium, or small drive sprockets. A scale, showing how to set machine for any drilling distance, is provided with each planter. If this scale is not available, a few minutes spent in experimenting with various settings will give the desired adjustment. The plates used for checking usually can be used for drilling, the operator using the shifting lever and sprocket adjustments to get desired spacing.

To test the accuracy of the drop and to determine if right seed plates are being used, jack up the planter, fill hoppers with seed, and turn the wheels. Trip check-forks by hand and catch the seed, keeping accurate check on each dropping. Planter should not be turned faster than 35 revolutions per minute.

To find out if planter is giving a good cross check, carefully dig up a row of at least eight hills crosswise, setting a stake

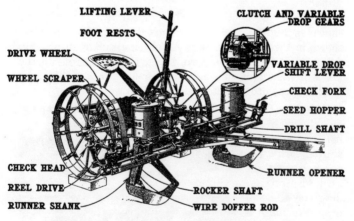

Figure 86—Two-row corn planter with important parts named.

in center of each hill. Due to the travel of wire, the hill of corn should be found about an inch behind the button. An adjustment is provided on planter which permits tilting front to place hills closer to or farther from button. Tilting front by lowering runner tips places hills farther back, while raising runner tip places hills closer to button.

To adjust width of the planter, remove bolts that hold shanks to the frame, remove bolt holding drive pinion on drive shaft, and adjust the shanks in or out to width of row desired. Be careful not to slip pinion off the shaft as the timing will be disturbed. Adjust wheels in line with the runners.

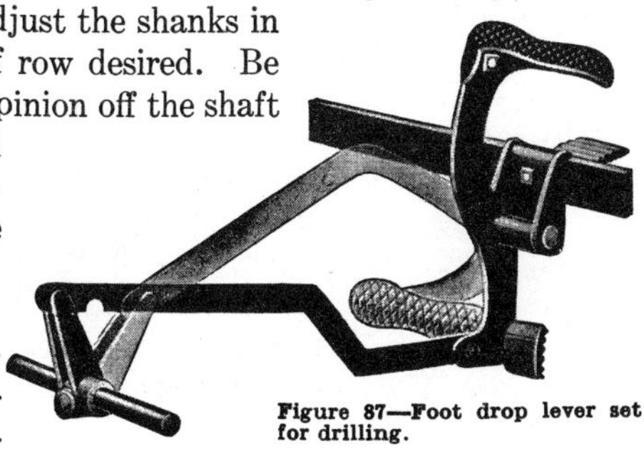

Field Operation. The object of check-rowing corn is to make cross cultivation possible. Cultivating crosswise of the rows is a difficult task if checking is not straight, and the straightness of crossrows depends more than anything else upon the handling of the check wire.

Figure 87—Foot drop lever set for drilling.

The check wire should be stretched reasonably tight when laid out and should be kept at that tension. The reel friction can be adjusted to hold wire to the desired tautness when unwinding, but uniform checking depends upon the judgment of the operator in pulling the wire to the same tension each time he moves the stakes.

Crooked crossrows may also be caused by running front of the planter at an improper level, in which case every pair of rows will be out

Figure 88—Two-row corn planter with fertilizer attachment.

of check. This may be adjusted as explained in preceding paragraph.

If one side only is out of check, it may be caused by valves not being adjusted correctly, frame of planter being bent, or by a weak rocker-shaft spring. A bent frame may also cause one row to be planted deeper than the other.

If the planter scatters seed between hills, the trouble may be due to kinks in the check wire, an obstruction in the valves, or too little tension on the rocker-shaft spring.

The four-row tractor-drawn planter, shown in Fig. 89, is similar in basic design to the horse-drawn planter described on preceding pages except, of course, that it has the bigger capacity desired by power farmers. The pressure wheel, set behind each runner, acts as an independent gauge wheel for each unit. As a result, each planting unit works independently of the others, thereby permitting each runner to ride over ridges or down into depressions without causing variation in planting depth of any of the other units. A power lift clutch on the planter (similar to that used on trac-tor plows) raises the planter runners and the disk marker

Figure 89—Four-row tractor planter checking corn on a mid-western farm.

at the same time. The instructions that apply to the horse-drawn planter, as far as seed plates and adjustments are concerned, are so nearly applicable to the tractor planter that a separate discussion on the tractor planter is not necessary for our purposes.

Caring for Corn Planters. To insure good work and accurate planting, corn planters must be well oiled and all parts must be firmly in position. Parts must be replaced when badly worn, or the efficiency of the planter is impaired.

Oil holes in a new planter should be filled with kerosene to cut out the paint, after which a good grade of machine oil should be used liberally on all friction parts with the exception of the enclosed clutch and variable drop gears which operate in a constant bath of oil inside the gear case. The gear case or housing in which these parts are enclosed should be filled to the level of the oil plug with clean, new oil of S.A.E. 20 viscosity. Inspect the oil level occasionally and if low, add sufficient new S.A.E. 20 oil to fill gear case to proper level. Before the planting season opens, remove drain plug from bottom of gear case, drain old oil, and flush out with kerosene. Refill housing to proper level with clean oil. Frequent oiling adds to the life of a planter except in extremely dusty conditions when it is better to use only kerosene on all working parts, excepting, of course, parts enclosed in housing.

When planting in wet or sticky soils, the operator should keep close watch to see that the seed boots do not become clogged with dirt which will stop the seed from reaching bottom of the furrow. It is a good plan to inspect the boots and the entire dropping mechanism at regular intervals to see that parts are working properly.

Equipment and Attachments. Several combinations of disk and runner furrow openers can be obtained to suit different soil conditions. Gauge wheels or gauge shoes can also be obtained.

In some sections of the country, fertilizer is sown when the corn is planted. A fertilizer attachment for this purpose is provided by most manufacturers, as shown on the planter illustrated in Fig. 88.

It is important that the fertilizer be kept from coming in direct contact with the seed. If this occurs, the seed is "fired." The fertilizer attachment, shown in Fig. 88, places a strip of fertilizer on each side of the hill after the seed has been partly covered with soil, so the fertilizer does not come in contact with the seed. Covering knives then throw soil over the fertilizer.

If it is desired to plant peas or beans along with the corn, the pea planting attachment may be added to the planter. Both the fertilizer and pea planting attachments can be used when planting corn, making it possible to plant two crops and sow fertilizer in one operation.

While planting accuracy depends almost entirely on the drop and the selection of seed plates, the tongue truck is an important factor in obtaining a more accurate check and better all-around results, especially in hilly land. It eliminates neck weight and up-and-down movement of the planter, reduces sideslip on hillsides, assures more uniform planting depth, better covering of the seed, permits driving the planter straighter, and assures easier turning at row ends. Planters can be furnished with tongue truck, or the tongue truck may be purchased as a separate unit and attached by the owner.

Figure 90—One-row corn drill.

Corn Drills. Fig. 90 shows a type of one-horse corn drill used in small farm districts. It is very simple to operate. The drop is the same as shown and described in connection with the two-row planter.

Questions

1. *What results when row crops are planted too thick? Too thin?*

2. *What is considered the proper number of kernels of corn to plant in a hill in your community? If corn is drilled, what distance is considered best?*

3. *What two factors determine the accuracy of a corn planter?*

4. *Name and describe the action of the two types of corn drops.*

5. *What test would you make in determining the size of seed plate to use?*

6. *How would you test the cross check of a corn planter?*

7. *What may cause crooked crossrows in checking and how can this be avoided?*

8. *What are the important points to remember in caring for a corn planter?*

9. *What attention should be given to the enclosed clutch and variable drop gears? How lubricated?*

10. *What special equipment or attachments are used on corn planters in your community?*

Chapter VI.

COTTON AND CORN PLANTERS

The cotton grower requires a planter that will plant cotton, corn, and other row crops with equal accuracy. His multi-purpose planter must be quickly and easily convertible from one type of planter to another.

There are several types of combination cotton and corn planters, each designed to meet planting conditions common to a certain section.

Fig. 91 illustrates the type of planter most generally used where cotton is raised on ridges. This machine can be obtained with either runner or shovel opener. Fig. 92 shows a type of two-row planter available for both drilling and checking both cotton and corn. One-row walking planters are also made for hill-dropping and drilling cotton and corn.

Figs. 93 and 94 show a four-row tractor planter used on an adjustable-tread general purpose type tractor. This

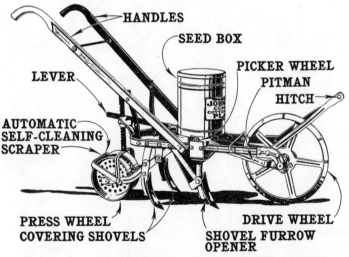

Figure 91—One-row combination cotton and corn planter with parts named.

equipment is gaining favor throughout the Cotton Belt because of its ability to cut production costs. While this planter is power-driven and power-lifted, its dropping mechanism is the same as that described for other cotton and corn planters.

The one-row planter, shown in Fig. 96, is especially adapted to planting in hard soils. It is of stronger build than the regular type of planter and will work well in the severest conditions. It is generally used with a sweep on the opener standard.

Planting Devices. Like corn planters, the success of combination cotton and corn planters depends upon the accuracy of the drop and its proper adjustment. The dropping device for all combination

Figure 92—Two-row combined cotton and corn planter.

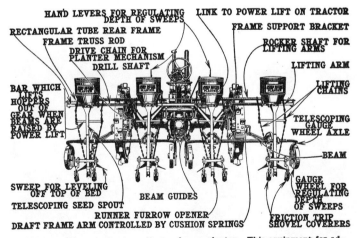

HAND LEVERS FOR REGULATING DEPTH OF SWEEPS   LINK TO POWER LIFT ON TRACTOR

RECTANGULAR TUBE REAR FRAME   FRAME SUPPORT BRACKET

FRAME TRUSS ROD

DRIVE CHAIN FOR PLANTER MECHANISM

ROCKER SHAFT FOR LIFTING ARMS

DRILL SHAFT

LIFTING ARM

BAR WHICH LIFTS HOPPERS OUT OF GEAR WHEN BEAMS ARE RAISED BY POWER LIFT

LIFTING CHAINS

TELESCOPING GAUGE WHEEL AXLE

BEAM

SWEEP FOR LEVELING OFF TOP OF BED   BEAM GUIDES

TELESCOPING SEED SPOUT

RUNNER FURROW OPENER

DRAFT FRAME ARM CONTROLLED BY CUSHION SPRINGS

GAUGE WHEEL FOR REGULATING DEPTH OF SWEEPS

FRICTION TRIP SHOVEL COVERERS

Figure 93—Four-row tractor cotton and corn planter. This equipment for adjustable-tread general purpose tractors is becoming very popular in the cotton-growing states.

cotton and corn planters shown are the same.

The corn drop is the same as that shown for the corn planter in Fig. 86. Distance of drilling is regulated by the number of cells in the seed plate in all except the two-row planter, shown in Fig. 92, which has, in addition, the variable drop control, described in the discussion on corn planters in the preceding chapter, and sprockets of various sizes for the drive chain. The selection of seed plates that suit the size of seed to be planted is one of the most important factors in getting an even stand of corn. Fig. 85 illustrates this point.

Cotton seed is one of the most difficult of all seeds to plant accurately. For that and other reasons, it is usually planted thicker than desired and the weaker plants chopped out. However, the grower wants as great a degree of accuracy as possible, and he needs to know how to adjust his planters to get best results. In unusually weedy conditions, many farmers prefer to check their cotton to permit cross-cultivation. Certain check-row planters give very good results even with undelinted cotton seed. Checking or hill-dropping cotton saves seed and reduces chopping.

Figure 94—Planting cotton with a four-row combination cotton and corn planter.

The saw-tooth type steel cotton picker wheel, shown in Fig. 95, picks out the cotton seeds, one or more at a time, taking lint and trash from the hopper with the seed, and plants any quantity per acre desired. Fig. 97 shows the hopper bottom with the cotton plate or spider in position. The spider revolves in an opposite direction to the picker wheel, and delivers the seed in position for the teeth of the wheel to pick it out.

Figure 95—Cotton picker wheel showing how wheel picks out cotton seed one at a time. Quantity planted per acre is controlled by turning thumb nut.

The cotton feed gate controls the amount of seed to be planted. Turning the thumb nut (see Fig. 95) to right or left increases or decreases the number of seeds picked out by the picker wheel.

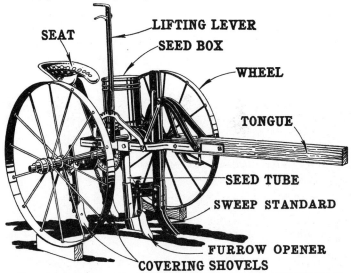

Figure 96—One-row combination cotton and corn planter especially adapted to planting on beds or in furrows. A sweep can be attached to heavy beam for sweeping off tops of beds when planting.

Changing Plates. To change from cotton to corn planting, remove thumb nut that holds the cotton spider, remove the spider, and insert the corn plate and cut-off. No other adjustment is needed. Reverse the procedure when changing from corn to cotton.

Practically all types of combination cotton and corn planters can be used for planting other row crops such as peanuts, kafir corn, etc., by using the proper-sized plates. Several types of planters are also adapted to sowing fertilizer at the time seed is planted.

Figure 97—Bottom of seed hopper showing cotton spider or plate in position.

Operation and Care. There are few field adjustments other than setting for proper planting depth. On the riding planters, levers provide this adjustment. On the walking planters, adjustment of the hitch or the press wheel, when used, gives desired depth.

Frequent oiling and an occasional general overhauling, with special attention to the tightening of all nuts, replacing worn parts, and testing the planting mechanism will add to the life and efficiency of cotton and corn planters.

Questions

1. *Describe several types of combination cotton and corn planters. With which type are you most familiar?*
2. *Why is cotton seed difficult to plant and how does the steel saw-tooth picker wheel overcome this difficulty?*
3. *How is the quantity of seed sown regulated?*
4. *How would you change from cotton to corn planting equipment?*

Chapter VII.
LISTERS

Listing is a method of raising corn or other row crops in regions having limited rainfall. Its advantages under these conditions are many. Planting the crops in the bottom of the furrow, then filling in the furrow by cultivating, keeps the plant roots deep below the surface where moisture is more plentiful. The preparation of the ground, previous to listing, is not expensive. Listed crops can be easily cultivated and kept free from weeds. For this reason, a farmer can care for a larger acreage of listed than of surface-planted row crops.

Single listing consists of planting the crop when the ground is listed the first time.

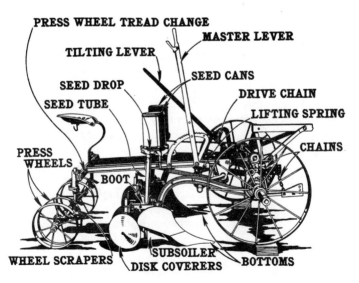

Figure 98—Two-row lister with important parts named.

Blank listing is listing the seed bed without planting, leaving the planting to be done with a regular corn planter or with the lister when double listing.

Double listing is the practice of blank listing in the fall to catch snow and hold the moisture, and splitting out the ridges and single listing in the spring.

Types of Listers.   Because of the fact that the lister combines the duties of the plow and the planter, its selection as to type is important.   Consideration should be given to style of bottom and coverers and to the size most practical for the acreage to be worked.

The type of lister illustrated in Fig. 98 is used in many listed-crop sections, and is probably the horse-drawn type in most general use.   It is adapted to general conditions and gives the farmer two-row working capacity.

The one-row lister, shown in Fig. 99, is representative of the type used in districts where rear press wheels are not desired.

The field scene, Fig. 101, shows a modern three-row tractor lister in operation.   This type of lister is gaining in popularity with farmers who work large acreages and farmers who are constantly endeavoring to lower production costs through the use of big-capacity machines.

There are other styles of listers than those mentioned above, but for the most part they are variations of these types.   The integral lister, illustrated in Fig. 100, is favored among owners of general-purpose tractors since it makes up a compact outfit with the tractor.   The power lift of the tractor, operated by the tractor engine, furnishes power for raising and lowering the lister.

Figure 99—One-row lister.

Bottoms and Planting Equipment.   The planting efficiency of a lister depends upon its bottoms and seed-dropping devices. The bottom, the subsoiler, and the coverers are responsible for the entire job of preparing the seed bed and covering the seed. The dropping devices determine the accuracy of planting and the spacing of the seed. When these two units are in proper adjustment, the operator can expect to do a good job of listing.

The duties of the lister bottom are similar to those of the plow bottom. It opens the seed bed, turning a furrow each way and pulverizing the soil in the same manner as does the plow. The quality of its work is even more important than in the case of the plow, as it is usually the only equipment used to prepare the soil for planting.

To do a good job, the lister bottom must be of proper shape, its share must be sharp and properly set, and its surface well polished for good scouring. When the share is sharpened, it is important that the wings are set alike. If not set the same, one side will cut deeper than the other and cause the lister to run to one side and pull heavier. To obtain proper penetration in all conditions, the share point

Figure 100—The integral lister makes up a compact plowing and planting outfit with the general-purpose tractor.

should have approximately the same amount of underpoint suction as a plowshare.

The planting mechanism used on the listers illustrated is the same as used on the corn planters and combination cotton and corn planters, described in preceding chapters. The corn drop is the same as that shown in Fig. 83. The sawtooth steel picker wheel, illustrated in Fig. 95, is used for cotton planting. Special plates for planting any row crop can be obtained.

The importance of selecting seed plates with cells of proper size to suit the seed to be planted cannot be over-emphasized. See Fig. 85, which shows how to select plate with proper-sized cells.

The seed plate and drive sprockets control the planting distance when planting crops other than cotton. The distance at which seed is planted varies with the number of cells in the seed plate used and the size of sprocket. The lister operator can select plates and sprockets to space the seed the desired distance in the drilled rows.

Figure 101—Double-listing with a three-row tractor-drawn lister.

Field Adjustments. The discussion of field adjustments and operation is based upon the lister shown in Fig. 98. The adjustment of other types may be somewhat different, but in general, the operation is the same.

On the two- and three-row listers, rows can be spaced at various distances apart by moving the hitch, bottoms, cans, and rear wheels in or out by means of the adjustments provided. The operator must be sure to move each the same distance so that all will be in line to plant rows of uniform width.

Depth is controlled by the master or depth lever. Bottoms are leveled to same depth with the tilting lever. Chains at front of beams are for leveling the bottoms in accordance with the depth. After depth has been set, chains should be lengthened or shortened so shares will have about 3/8-inch suction. If too much suction is given to the shares, the bottoms will kick up at the rear. If too little, the lister will not penetrate as it should.

Adjustment of the lifting spring is provided so the spring can be set at proper tension to aid in lifting the bottoms.

The subsoiler opens the seed trench. It should not be set deeper than necessary to do proper work—about one and one-half inches below the point of the share.

When rolling coulter is used, it must be set in line with the exact center of the bottom, or one furrow will be wider than the other, resulting in uneven work and side draft.

Covering equipment—disks or shovels—must be set alike, or it will lead the lister to one side.

Rear wheels must run straight. The lock casting on the upright angle below frame casting must be kept straight, or wheels cannot run parallel to each other. They can be staggered in or out to meet soil conditions by reversing the axle.

Care of listers. Length of service and working qualities of a lister depend greatly upon the care given it. All polished parts—bottoms, subsoilers, root cutters, and coverers should be coated with oil whenever the lister is not in use. Rust pits on these parts prevent scouring and hinder good work. Disk coverers, wheel boxings, and all points of friction should be oiled regularly.

General overhauling after each planting season with special attention given to worn and loose parts will add to the life of the lister.

Questions

1. *What is the purpose and what are the advantages of listing crops?*

2. *What is single listing? Blank listing? Double listing?*

3. *Describe several types of listers.*

4. *Why are the bottoms of first importance to good work in a lister?*

5. *How is planting distance controlled?*

6. *How is row-spacing controlled on two- and three-row listers?*

7. *What is the purpose of the master lever? Leveling lever? Beam chains?*

8. *What are the important points in caring for a lister?*

Chapter VIII.
POTATO PLANTERS

Growing potatoes for market without modern equipment for planting and harvesting them is an expensive and laborious task. The slow, difficult hand-drop method of planting has been supplanted to a great extent by mechanical planters that open a furrow, space the seed at the desired distance, and cover it at the proper depth. Mechanical diggers that remove and separate the potatoes from the soil and vines do away with the slow, tiresome practice of plowing out the crop with an ordinary plow.

The potato planter has come into quite common use in most sections of the country within the past thirty years. It has been an important factor in making potatoes a profitable crop—one that ranks high in value among the leading crops. Lower production costs due to saving time and labor and bigger yields because of uniform planting and covering are two advantages derived from the use of a potato planter.

Types of Planters. The type of potato planter in most common use is illustrated in Fig. 103. It is well adapted to

Figure 102—The modern potato planter not only speeds planting but insures the accuracy that means full yields.

planting conditions in practically every potato-growing section. A fertilizer attachment that deposits fertilizer on each side of the furrow and mixes it properly with the soil can be added, making a two-purpose machine that is both practical and economical.

When potatoes are raised on a large scale, two-row planters, of the type shown in Fig. 102, are used with a big saving in time and production costs. This machine combines big-capacity planting with the labor-saving advantages of the one-row planter. It, too, can be equipped with a fertilizer attachment, as shown.

Principle of Operation. Potato planters of the one-man, or picker type, shown in Fig. 103, have often been described as "almost human" in their work of picking out a piece of seed and dropping it in its proper place. Their work is probably more difficult than that of any other planting machine because of the irregularity in size and shape of the seed they are required to plant.

The picker bowl is shown in Fig. 104. Picker arms, revolving on the main axle, pass through concaves in a picking chamber containing the seed. Each picker arm is equipped

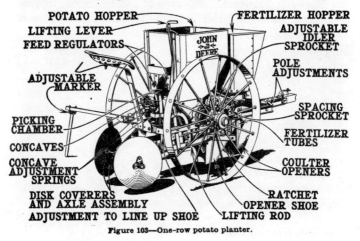

Figure 103—One-row potato planter.

with two sharp picking points which pick out one piece of seed. As the arm passes downward, the seed is forced off the picker points by contact of the picker arm head with a bumper plate. The seed drops down into the trench made by the furrow opener where it is covered by the disks at the rear.

Adjustments are provided to meet planting conditions encountered in any section of the country and under varying field and seed conditions.

Proper Feeding Necessary. One of the important requirements for accurate planting is the maintenance of the proper amount of seed in the picking chamber. For best results, it is recommended that not more than a hatful of potatoes be in the chamber at any time.

Agitator plates, one on each side of the hopper, are agitated by lugs cast into the hubs of the drive wheels. As these lugs come in contact with the plates, sufficient agitation is furnished to keep up a steady flow of seed into the picking chamber.

When the seed in the picking chamber reaches a given level, which provides about the right quantity of seed for efficient performance, no more seed will flow into the picking chamber except as required to replace seed planted.

The amount of agitation necessary to insure an even flow of seed from the hopper into the picking chamber, regardless of the size of potato seed used, is controlled by means of a feed regulator adjustment, as shown in Fig. 103. Once this adjustment is set to accommodate size of seed to be planted, no further adjustment of feed regulators is required.

Concaves. Each pocket in the concaves holds one piece of seed. As the pickers pass through the concaves, the seed-piece is pressed onto the picker points by the concaves. There are coil springs on the concaves; if this tension is too great, the seed is shoved too far onto the picker points; if too little, the points may fail to pick up a seed. Observation will tell the operator when the proper tension is obtained.

One lever controls both the furrow opener and disk coverers on the planter shown in Fig. 103. When these parts are lowered with the lever, the machine is automatically placed in gear. An auxiliary control is provided so that the picking device

Figure 104—Illustration showing picking operation. Seed-piece No. 1 is resting in the picker concave awaiting the arrival of the picker. Seed-piece No. 2 has been forced onto the picker points by the pressure of the spring-controlled concaves. Seed-guard at "A" drops against seed-piece to keep it in place. Seed-pieces No. 3 and No. 4 are approaching the top of their swing. In position No. 5, the seed-guard has dropped away from the seed-piece ready to be released. In position No. 6, the spring release, "B", trips the picker head and releases the seed to the trench.

can be thrown out of gear at the end of rows, while the covering disks remain in working position to cover the last few hills planted. The disk coverers are adjustable for covering at the desired depth.

Spacing of the seed in the row is dependent upon the size of sprocket wheel used on the intermediate shaft. Nine sizes of wheels provide thirteen different planting distances from 7 to 30 inches. Changing from one spacing to another is a simple operation.

Care Lengthens Life. Like all other machines, the potato planter will last longer and give more satisfactory service if it is oiled properly when in use and when stored. A thorough overhauling before each planting season will add to its efficiency and life.

Questions

1. *What are the advantages of using a potato planter?*
2. *What types of potato planters are used in your community?*
3. *Describe the principle of operation of the potato planter.*
4. *Why is proper feeding necessary and how is it controlled on the machine described?*
5. *What is the function of the concaves?*
6. *Describe, in general, the operation of a potato planter.*

Part Three
CULTIVATING

Proper cultivation of row crops during the growing season has much to do with the yield and the quality of the crop produced. This fact is apparent to anyone who has observed the results of good cultivation and of poor cultivation in adjacent fields. Corn, potatoes, or any other crop that is smothered by weeds will produce little compared with the well-cultivated crop.

The destruction of weeds is the primary purpose of cultivation. Weeds draw moisture and plant food from the soil, robbing the growing crop. Weeds are truly thieves in the fields; they steal profits when permitted to grow unhampered.

Cultivation serves two other purposes as well. It creates

Figure 105—Cultivating speeds up when the tractor cultivator takes over this important job.

a moisture-saving surface mulch and admits air and light to the soil.   If the soil becomes crusted after a rain, it should be broken into a mulch to prevent the escape of moisture through capillary attraction.   Air and light are essentials to plant growth and are admitted to the soil more readily when the surface is loose and ridged.

The first requirement of a cultivator is that it be quickly and easily adjustable to varying field conditions.   The operator will find it impossible to cut out all the weeds and stir the ground evenly unless his cultivator is adjusted properly.   Angle of the shovels, tilt of the rig-beams, depth of each rig, and setting of pole must be correct for efficient work. To be easily adaptable to all conditions is the most valuable attribute of a cultivator.

Types of Cultivators.   Although not strictly a cultivator, the rotary hoe must be classed as a cultivator because of the work it does.   Its main purpose is to destroy weeds and create a surface mulch.

When classified according to capacity and mode of operation, there are one-row walking, one-, two-, and three-row riding horse-drawn, and two-, three-, four-, and five-row tractor-drawn cultivators.

Classified according to equipment used, and the conditions for which they are designed, there are shovel, disk, and listed-crop cultivators.   This classification will be used in the following text.

Chapter IX.

ROTARY HOES

The rotary hoe has proved itself of considerable value when used in the early stages of crop growth, this being especially true in corn and other row crops.   It is also used with success in small grain crops and in any condition where it is desired to break up a crusted surface soil. It is used to best advantage after heavy rains have packed the ground and created an unsatisfactory surface tilth.

Hoe Teeth Stir Soil. The rotary hoe is made up of two series of hoe wheels, one series mounted on the front gang axle and the other on the rear gang axle, spaced so the rear wheels work the soil left between the front wheels. Each hoe wheel has 16 teeth, shaped like fingers, which penetrate and stir the soil as the wheels rotate. A thoroughly pulverized surface results. Weeds are uprooted and the soil is left in good condition for plant growth.

While its work is most satisfactory when crops are just coming through the ground, the rotary hoe is used to good advantage in stimulating the growth of crops after they have passed this stage. Some farmers report satisfactory results using a rotary hoe in corn that has reached a height of ten inches.

Two sizes of rotary hoes are in general use—the two-row and the three-row shown in Fig. 107. The three-row is probably more economical for the average farmer because

Figure 106—A hook-up of two two-row rotary hoes. This outfit works a strip 14 feet wide each trip across the field. Hitch is flexible, allowing the two units to move up and down on uneven ground and work all the soil.

of its greater capacity. On large farms, where extra capacity is desired, the practice of operating special hook-ups of two or three rotary hoes behind the tractor is growing in popularity. (See Fig. 106.) Special hitches are furnished by the manufacturer for this purpose.

Easy to Operate. The rotary hoe is easy to operate and adjust. Once the machine has been set at the proper depth, the operator has little to do but drive his team or tractor.

Questions

1. *Describe the action of a rotary hoe.*

2. *Under what conditions do rotary hoes do the best work?*

3. *What crops are cultivated with rotary hoes in your community?*

4. *What sizes of rotary hoes are in common use in your community and which is most popular? Why?*

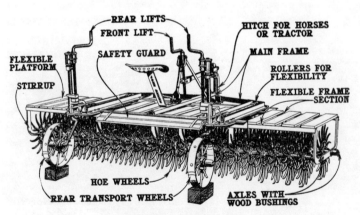

Figure 107—Three-row rotary hoe.

Chapter X.
SHOVEL CULTIVATORS

Shovel cultivators are the most generally used type in the majority of farming communities. They are suited to practically all soils and produce most satisfactory results when used in average conditions.

Fig. 108 illustrates a popular style of two-row cultivator. It is a typical two-row for use with three or four horses. Levers are provided to make all field adjustments. The master lever raises all four rigs. Independent depth levers raise or lower each rig separately, permitting adjustment for good work in uneven ground. Spacing lever moves shovels closer to or farther from the rows, giving operator instant control of spacing while cultivator is at work. The tilting lever levels rigs, causing front and rear shovels to run at same depth when going up or down hill or on uneven ground.

The field operation and adjustment of two-row cultivators is discussed in a later paragraph along with that of all shovel cultivators.

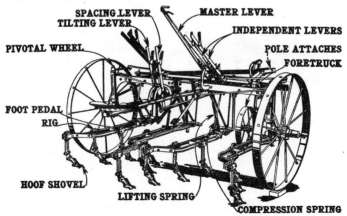

Figure 108—Two-row shovel cultivator. This machine can be equipped with any one of the combinations of shovel equipment shown in Figure 112.

One-Row Cultivators. One-row shovel cultivators are built in both walking and riding styles. Farmers in some sections, where farms are generally small, use the walking type of cultivator. Farmers in other sections much prefer the riding type.

Fig. 109 shows a style of walking cultivator in common use. This style is easy to handle and adjust for good work. It is especially popular in southern and eastern sections.

The type of one-row cultivator illustrated in Fig. 110 is probably the most widely used shovel cultivator. It is shown with hoof shovels and pin-break rigs, equipment used quite generally. Other beam and shovel equipment that can be used on this and other styles of shovel cultivators is shown in Fig. 112.

These different combinations of equipment are necessary to meet the requirements of each section of the country and their use is governed by soil conditions and crops grown.

The lever adjustments on the one-row (Fig. 110) are the

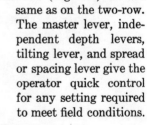

Figure 109—One-row walking cultivator.

same as on the two-row. The master lever, independent depth levers, tilting lever, and spread or spacing lever give the operator quick control for any setting required to meet field conditions.

The swinging rig type of cultivator, shown in Fig. 109, is adjusted for spacing at front end of the rigs. For cultivating closer to or farther from the row, loosen the arches and slide them in or out to width desired.

Operation and Adjustment. When a cultivator is set properly for working in ordinary conditions, the shovels will penetrate well, run at the same depth without crowding toward or away from the row, and there will be no unnecessary draft. To set and maintain a cultivator in this desirable adjustment is comparatively easy, once the operator understands the causes of trouble and the adjustments provided on his cultivator for correcting them.

One of the most common adjustments that cultivator operators have to make is setting the shovels for proper penetration and uniform depth.   The first requirement of an efficient shovel is that it be sharp, with the point properly shaped. A dull shovel will not penetrate easily; it does poor work and causes heavier draft.   Fig. 111 shows a properly shaped shovel with a dotted line showing its shape when point is worn and dull.   Frequent sharpening of shovels will insure a smooth-running, good-working cultivator.

All shovels must run at the same depth for good work. If the front shovels run deeper than the rear ones, all shovels

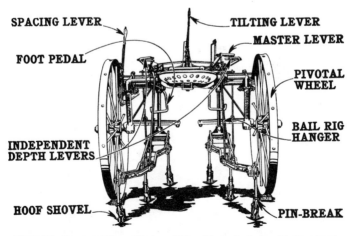

SPACING LEVER

FOOT PEDAL

TILTING LEVER

MASTER LEVER

PIVOTAL WHEEL

BAIL RIG HANGER

INDEPENDENT DEPTH LEVERS

HOOF SHOVEL

PIN-BREAK

Figure 110—One-row riding cultivator.   This cultivator is shown with other types of equipment in Figure 112.

will stand straighter than they should, and will not penetrate easily. This condition is due to the front of rigs being lower than the rear. It can be corrected by raising the pole at hames or leveling the rigs with the tilting or leveling lever.

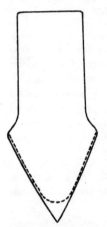

If the rear shovels run deeper than those in front, all shovels set too flat and will not penetrate as they should. This can be remedied with the tilting lever on some types of cultivators and with adjustment at the hames on types that do not have such a lever. These directions are based upon the supposition, of course, that all shovels are at uniform height on the shanks.

Pitch of Shovels. All cultivators are provided with an adjustment, either on the shovel shank or sleeve, whereby the pitch or angle of the shovel can be changed. This adjustment is correct for average soil conditions when the cultivator leaves the factory. However, it may become changed and it is well to know what the proper pitch is and how to get it when conditions demand.

Figure 111—Cultivator shovel with dotted lines showing how point looks when shovel needs resharpening. Obviously, a shovel in this condition penetrates poorly and increases draft.

Figure 112—Three types of cultivator rig equipment. Left, eight-shovel, I-beam rigs with spring-trip shanks; center, pipe beams with spring trips; right, spring-tooth rigs.

If a shovel stands too straight, it will not penetrate readily; it will not run steadily. There will be a tendency to skip and jump, and it will require unusual pressure to keep the shovels at work. If set too flat, the underpart of the shovel will ride below the extreme point and the shovel will not penetrate unless forced into the ground.

The illustrations in Fig. 113 show the proper pitch of shovels for good work compared with shovels set too straight and too slanting.

In hilling row crops, it is necessary to turn the front shovels in by loosening the clamp attachment on the shank. This setting tends to pull the shovels away from the row. This tendency does not interfere with the work of a pivot axle cultivator, but with the swinging rig type, the operator finds diffi-

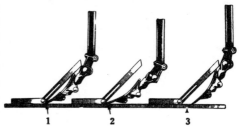

Figure 113—Showing correct and incorrect pitch of cultivator shovels. No. 1, shovel properly adjusted, No. 2, shovel set too flat, will not penetrate well; No. 3 shovel set too straight, will not penetrate or run steadily.

culty in keeping the rigs running the proper distance from the row when spread arch is not used. An opposite effect is produced when the front shovels are turned away from the row for first cultivation. In either case, the crowding tendency can be overcome and the rigs made to run straight by turning the rear shovel to an equal angle in the opposite direction. On parallel rig types of cultivators, turning the shovels in or out does not affect the operation of the cultivator.

The wheel tread is adjustable on most types of cultivators. This is an important feature in districts where several widths of rows must be cultivated with the same machines. The one-row cultivators, shown in Figs. 109 and 110, can be adjusted to several different row widths by removing a cotter

key or loosening a set screw and moving axles in or out an
equal distance to the desired position. Other types of cultivators are adjusted in a similar manner.

Adjustment of Shields. Proper setting of the shields
is important to good work during the first cultivation.
There are three main types used—solid sheet iron, open rod
wire, and rotating shield. The first, which is the most commonly used, is comparatively easy to set to allow the desired
amount of dirt to roll up to the row without covering the
plants. The rotating shield is often set too far back to
be efficient. The greater part of the shield should be ahead
on the front shovel, with its entire weight resting on the
ground. For later cultivation, shields are removed, provided
the crop has reached sufficient height.

Care of Cultivators. Like all other farm implements, the
length of life and the satisfaction given by shovel cultivators
depend upon the way they are handled and the care given
them during operation and storage. A few minutes given to
inspection and tightening of all parts, and thorough oiling at

Figure 114—Corn Belt farmer doing good work with a two-row cultivator.

regular intervals while in the field will add to the service of a cultivator and save delays caused by breakage and wear. Shovels should be polished and coated with oil when standing overnight, and covered thoroughly with heavy grease when stored.

As stated previously, one of the most important factors in efficient cultivation is keeping the shovels sharp. A dull shovel is as inefficient as a dull knife. It is advisable to have the shovels sharpened and shaped by a good blacksmith during the storage season. If the points are too badly worn, new shovels or new points (on the slip-point type of shovel) should be obtained. If the shovels have become rusted and pitted between seasons, they should be polished before being taken into the field.

The slack season is the time to go over the cultivator thoroughly, ordering new parts wherever needed and tuning it up ready for the first day of the cultivating season.

Questions

1. *What are the purposes of cultivation?*

2. *What are the requirements of a good cultivator?*

3. *Name several types of shovel cultivators. Which type is most generally used in your community?*

4. *Describe the essential lever adjustments on a two-row cultivator.*

5. *What conditions govern the type of rig equipment used? What type is used in your community?*

6. *Why is it important that cultivator shovels be kept sharp?*

7. *What causes front shovels to run deeper than rear ones? How corrected?*

8. *What relation has the pitch of a cultivator shovel to its work?*

9. *How would you set shovels to cause the rigs to run straight when hilling?*

10. *What is the advantage of an adjustable wheel tread?*

11. *Of what value are shields and how are they adjusted?*

12. *Tell how you would care for a shovel cultivator when in use and in storage.*

Tractor Cultivators

Its adaptability to cultivating row crops is one of the chief reasons for the fast-growing popularity of the general purpose type of tractor. Field experience has proved that tractor cultivators effect great savings in time and labor. The steady speed of a tractor cannot be matched by horses, especially on hot days. And, too, the speed can be controlled to meet crop and field conditions to best advantage.

Owners of tractor cultivators find they can put in more hours per day in the field—cultivating capacity is not limited to the endurance of animal power. In rush seasons, cultivating can be finished sooner, and the time saved can be utilized in taking care of other crops. When the weather is unsettled, the owner of a tractor cultivator can take full advantage of favorable conditions.

Operation and Care. The two-row tractor cultivator, shown in Fig. 115, is adapted to use in crops planted with two- or four-row tractor planters. The four-row cultivator, illustrated in Fig. 116, is used in cultivating row crops planted with four-row planters. These cultivators form a single unit with the tractors and are controlled entirely from the tractor

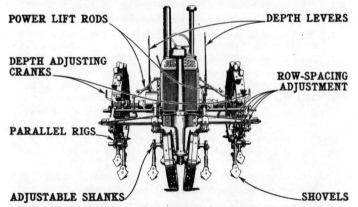

POWER LIFT RODS · DEPTH LEVERS

DEPTH ADJUSTING CRANKS

ROW-SPACING ADJUSTMENT

PARALLEL RIGS

ADJUSTABLE SHANKS · SHOVELS

Figure 115—Two-row tractor cultivator mounted on general purpose adjustable-tread type of tractor.

seat. The entire unit steers with the tractor—once the equipment is properly set, the operator's only duty is steering the tractor and setting the power lift into action at the row ends.

The pipe frame on the four-row cultivator, shown in Fig. 116, is rigid, flexibility being obtained through the floating construction of the rigs. The depth of cultivating is automatically controlled by individual gauge wheels, the rigs following the contour of the field as the tractor moves along.

When the operator reaches the end of the field with either of these cultivators, he touches a pedal which sets the power lift into action. The power lift raises all of the rigs—the turn is made without stopping—the operator touches the pedal again and the rigs are lowered to work.

The shovels used on the tractor cultivators shown here are practically the same as those described in connection with the horse-drawn cultivators on preceding pages. They should be adjusted and cared for in the same manner as

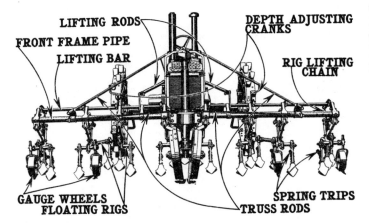

Figure 116—Four-row tractor cultivator on a general purpose adjustable-tread tractor. This big-capacity type outfit is becoming more popular each year.

Figure 117—Two-row tractor cultivator equipped with spring-tooth rigs.

other cultivator shovels. Spring trips protect the shovels and shanks against breakage. On the two-row cultivator shown in Fig. 117 the spring teeth act as the safety factor to protect cultivator from breakage.

Careful adjustment and thorough oiling of cultivating equipment will increase the efficiency and lengthen the life of tractor cultivators.

Questions

1. *What are some of the advantages of owning a tractor cultivator?*

2. *How is the cultivator guided?*

3. *Why is it necessary that a tractor cultivator have flexibility?*

4. *How is the cultivating equipment, on the cultivators illustrated, raised and lowered?*

5. *How would you adjust depth of each rig individually or all rigs at one time?*

Chapter XI.
DISK CULTIVATORS

Under certain conditions, disk cultivators are used to better advantage than shovel cultivators. Their use is not general or widespread, being localized in many sections of the South, East, and Central States.

Disk cultivators are adapted to many different conditions. They are used in unusually weedy fields where shovel or surface cultivators would have a tendency to clog. In stony or rooty fields, and for hilling crops, the disk type of cultivator is favored.

Figure 118—Doing good work on crooked rows with a disk cultivator.

A favored type of disk cultivator that can be converted into a shovel or spring-tooth cultivator by simply changing the rigs is shown at work in Fig. 118. This type of machine is often referred to as "three-in-one" cultivator because of its convertibility, giving the owner three types of cultivators for the cost of one, plus the cost of the two additional rig equipments.

Disk cultivators are operated in much the same manner as shovel cultivators. A master lever raises both rigs; independent levers for each rig control depth and give adjustment for varying field conditions. A crank at rear of the pole controls leveling of the rigs.

To change disks from out-throw to in-throw, the rigs are reversed, left to right. This is done by removing two cotter keys and tipping the gang so that the dog will clear the upright ratchet. The gangs are then transferred. The change from disk to shovel or spring-tooth rig equipment is made in the same manner.

Two ratchets with grip levers provide adjustment for angling and tilting the gangs to get desired results with any equipment.

Wheels, disks, and all other parts should be well oiled.

Questions

1. *Describe the uses of the disk cultivator.*

2. *What are the advantages of a three-in-one disk cultivator?*

3. *How would you change a disk cultivator from in-throw to out-throw?*

Chapter XII.

LISTED CROP CULTIVATORS

The fact that listed crops are planted in trenches makes necessary the use of special cultivators for first and second cultivations. Cultivators designed for cultivating surface-planted crops will not hold to the ridges and cut the edges of the trenches properly the first time through. After two cultivations with listed crop cultivators, the field is leveled sufficiently for the use of regular cultivators.

The chief requirement of a listed crop cultivator is that it be quickly and easily adaptable to field conditions. Its equipment must be quickly adjustable to different width of rows, to trenches of various widths and depths, and it must be flexible enough to hug the ridges where rows are not exactly parallel.

Types. The two-row cultivator, shown in Fig. 119, is the common type of horse-drawn listed crop cultivator. It is

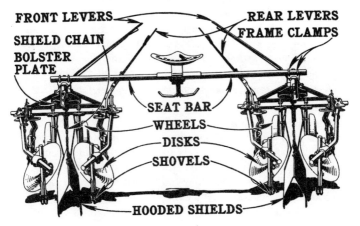

Figure 119—Two-row listed crop cultivator set for first cultivation.

shown equipped and set for first cultivation.   The ease with which it can be adjusted, its flexibility, and its two-row capacity have made this type of horse-drawn cultivator popular in listed crops.

Three-row horse-drawn cultivators, similar in construction and operation to the two-row, are available for listed crops. With the exception of a third gang, which is attached to the frame bar behind the operator's seat, they are practically the same as the machine shown in Fig. 119.   The three-row gives the listed-crop farmer 50 per cent greater working capacity than with the two-row, using two additional horses.

Three-, four-, and five-row tractor-drawn listed crop cultivators are in general use in listed crop territories.   With a cultivator of this type, one man can cultivate from 25 to 65 acres or more per day with approximately the same effort that is required to operate a machine of smaller capacity. Production costs are reduced and more time is available for the handling of other crops when such machines are used.

The general purpose tractor, gaining wide acceptance as cost-reducing power in listed-crop territories, has brought on the integral listed crop cultivator, shown in Fig. 120.   In addition, several types of tractor shovel cultivators are available with listed crop attachments for use in working down the ridges in the first two cultivations.   Thereafter, the listed crop attachment is removed and the cultivator used as a shovel cultivator for subsequent cultivations.

A great variety of disk, shovel, sweep, and surface cultivating equipment is available for use with listed crop cultivators.

Operation and Adjustment.   Listed crop cultivators are easier to operate and have fewer adjustments than shovel cultivators.   Once the shovels, disks, and shields or other equipment are set properly, the operator simply drives straight down the rows.   The only work required is raising and lowering the equipment with the levers provided.

To get the proper setting of cultivating equipment, the operator must take his machine into the field and adjust it to the particular job to be done.   For first cultivation, the disks are set to cut the weeds on the sides of the trenches, throwing away from the row.   The shovels are set to stir the edge of the trench and, if four shovels are used on each gang, to cultivate between the rows.   The hooded shield is run on the ground to prevent dirt from rolling in on the young plants.

When cultivating the second time, disks are set to roll dirt into the trench, around the growing crop, and shovels are set wide to cultivate between the rows.   It is usually desired to leave the field as level as possible the second time through, so that regular cultivators can be used in place of the listed crop type in future cultivations.

All adjustments of equipment are made with clamps. The gangs are spaced on the seat bar with the frame clamps for different row spacings.   Disks can be set to throw in or out, working from five to thirty-six inches apart, with the clamps provided.   Shovels can be set at any angle, ten to fifty-four inches apart, with clamps.   The wheels can be set to conform to width of the trenches with a clamp adjust-

Figure 120—Two-row integral listed crop cultivator at work with a general-purpose tractor.

ment. Simply loosening the clamps permits the shifting of this equipment to the desired position.

The shovels and disks of the horse-drawn styles are raised at the ends of the rows with the front levers, one to each gang. The rear levers control the shovels only, permitting the operator to change depth or shake out trash. A coil spring with adjustable tension aids in the operation of the rear levers.

One master lever raises shovels and disks of all gangs on the tractor-drawn cultivators. It is within reach of the operator on the tractor seat and is operated without stopping the outfit. Two auxiliary levers on the forward frame lift the outside gangs independently of the master lever. On each gang is a depth lever which is used, also, for shaking out trash.

Oiling Important. Repair costs are lessened and length of life increased by regular oiling and greasing of all working parts of a cultivator. Disk and wheel bearings are provided with hard oil cups on the end of the spindle. To lubricate the bearing, unscrew cup, fill with hard oil and screw cup up tight to place and screw down the set screw. Bearings should be greased once each day. All moving parts should be oiled frequently.

An occasional thorough inspection of the machine may save the operator time and trouble. A cultivator is easier to operate and less breakage occurs if all nuts are kept tight and cotter keys split. Worn parts should be replaced by new parts before they cause delays through breakage at cultivating time. Winter months usually afford time for overhauling, replacement of worn parts, and general conditioning of cultivators for the coming season.

Questions

1. *Why is a listed crop cultivator necessary in cultivating listed crops the first and second times?*
2. *Name the essentials of a good listed crop cultivator.*
3. *Describe the proper adjustment of cultivating equipment for both first and second cultivations.*
4. *What is the purpose of the master lever?*
5. *How would you care for a listed crop cultivator during operation and between seasons?*

Part Four

HARVESTING

Harvest season is the busiest time of the year on every farm. All hands are turned to gathering the returns of the year's work. Grain, hay, food crops, and fruits must be harvested at the proper time if losses are to be avoided.

The small grain is ready to cut; a delay of a few days may cause heavy losses. Hay must be cut, cured, and stored in the shortest possible time to retain its maximum feeding value. Fruits are ripe and must be picked for the market. Then, all forces should work in harmony at the harvest.

Then, too, farm machines must be at their highest state of efficiency. Binders, mowers, rakes, pickers, combines, threshers, and all other machines must be tuned up in advance for speed and good work. Delays must be avoided. The operator must know how to make adjustments and repairs to get the most from his machines. He should be familiar with causes of inefficiency and know how to correct them with a minimum of effort and time.

Harvesting machines are probably the most intricate of all to adjust and operate. However, once the operator knows his machine and the adjustments that must be made most often, efficiency is maintained with little difficulty. The most common causes of mower and binder trouble can be avoided by thorough and regular oiling, by replacing worn parts before they affect the machine's operation, and by an occasional complete overhauling and adjustment of parts.

Chapter XIII.
GRAIN BINDERS

When it is time to harvest small grains, speed and good work are necessary to save all of the crop at the lowest possible cost. Binders must be at a high state of efficiency to handle the cutting and binding with few stops and few missed bundles. The operator must know how to adjust his binder to obtain the best results and complete the harvest most economically.

Since the introduction of the reaper and the first self-binder, steady advancement has been made in binder design. The binders on the market today, for the most part, are built to a high degree of perfection in the essentials necessary to good harvesting.

Types of Binders. Horse-drawn grain binders are in more general use today than are tractor-drawn machines.

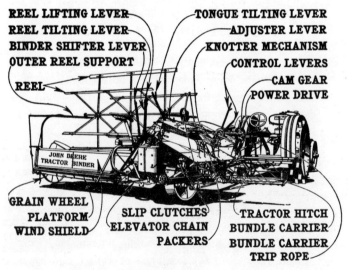

Figure 121—Tractor binder with tractor attached.

However, with the rapid increase in the use of tractors for power, it is reasonable to assume that before many years the tractor binder will equal or surpass the horse-drawn machines in numbers used.

The mechanism of tractor binders is driven direct from the tractor by power transmitted through a power drive shaft (see Fig. 121). The main or bull wheel of the binder merely carries the weight. This results in steady, even operation of the binder in all field conditions—wet ground, loose ground, down or tangled grain do not cause delays or interfere with good work as might be the case with horse-drawn binders.

In unusually heavy grain, the forward speed of the tractor can be reduced while the speed of the binder mechanism is maintained. Thus, heavy crops can be handled without clogging or stopping.

Levers that can be handled from the tractor seat are obtainable for most tractor binders. This permits one man to operate both the tractor and binder. Equipment for controlling the tractor from the binder seat can also be obtained.

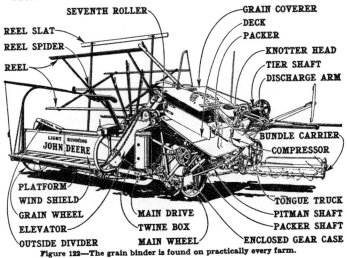

Figure 122—The grain binder is found on practically every farm.

The general operating instructions for tractor binders are much the same as for horse-drawn machines. The discussion of all grain binders will center around the horse-drawn machine, shown in Fig. 122.

For harvesting rice, a special binder is needed because of the wet and ridged condition that usually exists in the fields at harvest time. Fig. 124 shows a rice binder, the operation and care of which are essentially the same as of the horse-drawn grain binder. This binder is operated by tractor power, but can be obtained for use with horses as well.

Oiling of First Importance. There is probably no other farm machine that requires more thorough and regular oiling than the grain binder. Its draft, efficiency, and length of life depend greatly upon the liberal use of good, heavy machine oil on all bearings and working parts.

Manufacturers are constantly striving to build their binders with oil holes, cups, plugs, and grease fittings more accessible to the operator. The binders shown in Figs. 121 and 122 are equipped with a high-pressure grease gun oiling system, making thorough oiling an easy job.

The main drive gears, which transfer power from the bull wheel for operating the horse-drawn binder, shown in Fig. 122,

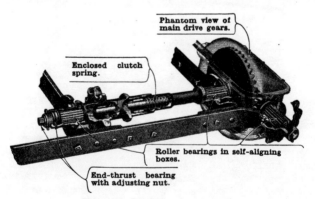

Figure 123—Detail of main drive gears.

are completely encased and operate in grease, as shown in Fig. 123. An occasional check to maintain level of grease in the housing is all that is necessary to keep the main drive gears lubricated properly.

Setting Levers for Good Work. Efficient operation of a grain binder requires the proper setting of all levers to meet all field conditions to best advantage. Poorly shaped bundles, scattering of grain, and heavy draft are often caused by failing to make proper use of the levers. There is usually a real necessity for making frequent lever adjustments in even the best of field conditions.

The raising and lowering devices, on the main and grain wheels, that set the platform higher or lower, together with the tilting lever and the adjustable tongue tilting connection straps, provide a wide range of adjustment for cutting stubble high or low, and for tilting the platform to the necessary angle. These adjustments are found on all binders, yet they are used to best advantage by few binder operators.

The tilting lever controls the forward and backward tilt of the binder. For ordinary cutting, it should be set so that the binder platform tilts slightly forward. When cutting lodged or tangled grain, the machine should be tilted low to get as many of the heads as possible. If it is necessary to tilt the platform in this manner, the binder should be raised a little more than halfway on the main wheel and grain wheel hangers.

The purpose of the reel is to tip the cut grain onto the platform canvas as evenly as possible, in position to make

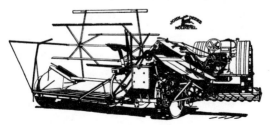

Figure 124—Tractor-drawn rice binder.

neat bundles. Reel levers provide a wide range of adjustment for setting the reel for good work. Under normal conditions, the reel slats should touch the grain close to the heads and leave the grain just after it is cut off. If there is a strong wind, it will be necessary to make test adjustments to see in which position the reel will do the best job of placing the grain evenly on the canvas. In lodged grain, the reel is set low and forward to aid in getting more of the down heads and straw.

The adjuster lever controls the position of the butter. The butter should be run as far forward as possible so that the grain is not forced backward from its line of travel as it comes up between the elevator canvases. In very short grain, it may be necessary to shift the adjuster back in order to place the band near the center of the bundle. Otherwise, the adjuster lever should not be used for regulating placing of the band.

Shifting the binder attachment with the binder shifter lever regulates the position of the twine on the bundle. It should be placed near the center of the bundle in grain of any height. It may be necessary to shift the binder with the binder shifter lever several times in making a round, if the grain is of uneven height. This adjustment should be made as often as necessary to make uniform, well-tied bundles.

The platform grain shield and the windboard at rear of the binder deck will assist in making better bundles and prevent scattering of grain, if properly adjusted for long or short straw.

Adjustment of Chains, Canvases. For best results, all chains should be run just tight enough to do the work. Draft and wear are increased materially by having them too tight. If the chains are run too loose, they will climb the sprockets, break links, and wear the chain or sprockets out rapidly. A loose main wheel chain is very often the cause of the binder choking.

All of the chains on a grain binder are provided with tighteners that can be quickly adjusted to take up slack or relieve the strain of an overtight condition. The binder operator will get more service from chains that are properly adjusted for good work and light draft.

An occasional oiling lengthens the life of chains.

Fig. 125 illustrates the proper method of attaching a chain on a sprocket and the best way to detach a chain.

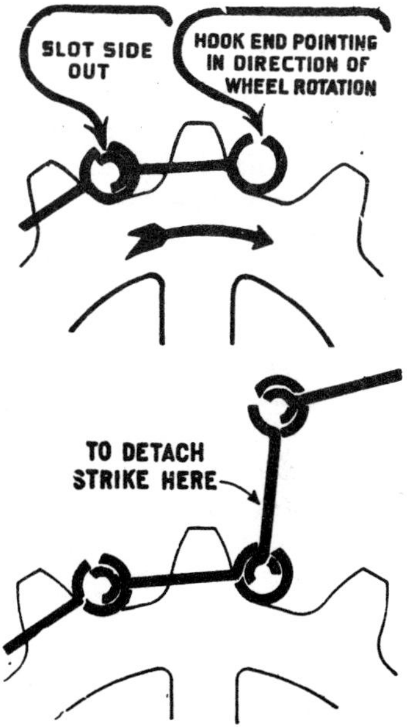

Canvases should be run just tight enough to do their work of carrying the grain to the binding unit. If run too tight, wear and

Figure 125—For best results, put chains on sprockets, as shown in the upper illustration. Lower illustration shows how to detach a chain; using a tooth of the wheel for a brace, bend to the position shown, and strike the link at the point indicated by arrow.

draft are increased materially. Two tighteners, operated by convenient levers, are provided for tightening the canvases on the binder shown. When a canvas is put on, all straps must be adjusted the same or the canvas will not run straight. The roller bearings on the elevator rollers that carry the canvases should be oiled regularly, as a sluggish roller increases draft and may cause clogging.

Binding Attachment Adjustments. The binding attachment is properly adjusted when the binder leaves the factory, and will operate successfully under average conditions without adjusting. It is a good plan to make no such adjustments on a new binder until the paint is worn off the working parts and they become smooth. However, when it becomes apparent that an adjustment will be necessary to insure efficient work, the operator should determine where the trouble exists before tampering with the parts of the binding attachment. If knotter or twine tension adjustments are made and do not correct the trouble, they should be changed back to their original position.

A few drops of oil will oftentimes correct a difficulty that appears to be serious. A loose nut may cause troubles beyond estimate. The shrewd binder operator will look for small troubles first, before attempting to make major adjustments of the binding attachment.

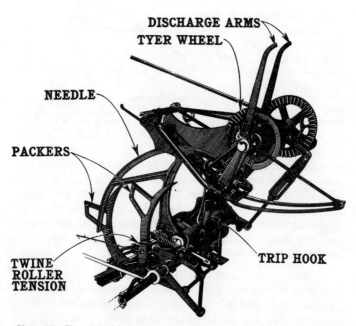

Figure 126—View of the binding unit with some of the parts named and indicated by arrows.

The illustrations in Figs. 127, 128, 129, and 130 show how the knot is tied. Careful study of this process will aid in a better understanding of the principles of binder operation.

Twine Tension. Incorrect twine tension is one of the most common causes of missed bundles. The twine should furnish a resistance of from six to eight pounds at the point of the needle. This is necessary to prevent slack interfering with the operation of the tying parts.

Twine tension is determined in the following manner: thread the machine with the exception of bringing the needle over to get the twine in the disk. Tie a loop in the twine under the breastplate, hook a fifty-pound spring balance or draw scale into the loop and pull the twine through the needle eye, pull on a line parallel with the binder deck. The scale should register between six and eight pounds if proper tension has been created.

The twine roller spring under the binder deck (see Fig. 126) controls the tension of the twine and is adjusted by loosening or tightening the tension spring. The spring should have free action and the rollers should turn freely with the pull of the twine if the tension test is to be accurate. This adjustment does not affect the tension of the band around the bundle. The trip stop spring affords adjustment for this purpose.

Trip Tension. To determine the number of pounds required to trip the binder and start the tying process, hook a scale into the upper end of the trip hook (see Fig. 126) and pull in a line parallel to the discharge arms when they have completed the discharge of a bundle. The tension should be twenty to twenty-two pounds.

The tension of the trip hook determines the tension of the twine around the bundle. If it is desired to increase or decrease this tension, the trip stop spring (Fig. 131) is adjusted accordingly.

Size of bundle is controlled by adjusting the trip hook in or

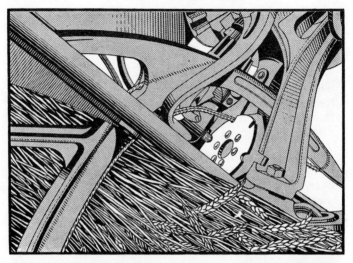

Figure 127—Binder has been tripped and needle has advanced, bringing twine around bundle and placing second strand over knotter jaw and into disk notch.

Figure 128—Disk has advanced and grasped both strands of twine. Knotter hook has turned, forming a loop of twine around the hook, and jaw has opened to receive twine leading to disk.

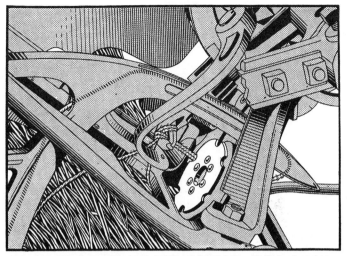

Figure 129—Knotter jaw has closed, holding twine tightly.   Knife is advancing, ready to cut twine between knotter hook and disk.

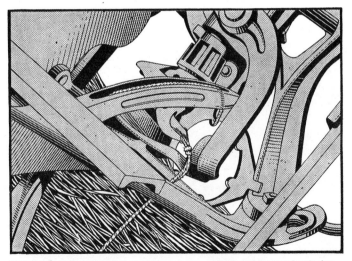

Figure 130—Twine has been cut; discharge arms, which have been turning during tying operation, strike bundle as twine is cut, assisting knife arm to strip loop from knotter hook.   Knotter jaw holds ends while loop is slipped off, making complete and secure knot.   With end of twine in disk, needle returns to home position, leaving strand of twine over knotter jaw.

out on the trip hook arm.   If small bundles are desired, the trip hook is set closer to the needle and the trip stop spring is loosened slightly.   If large bundles are wanted, the trip hook is set out on the trip arm and the trip stop spring is tightened.

Twine Holder or Disk Tension.   Perfect tying of knots is possible only when the twine holder or disk is set at the proper tension—thirty-five to forty pounds.   To test this tension, thread the machine ready for tying, pull the trip hook, and turn the discharge arms one complete revolution.   Take off the band and knot that were completed so as to have the twine securely in the disk.   Tie a loop in the twine directly above the knotter frame, hook the scale into this loop and pull straight upward (Fig. 132).   By tightening or loosening the twine holder spring (Fig. 133), the required tension is obtained.

When adjusting the twine holder spring, it is necessary to loosen the lock nut before the set screw can be turned.   It is advisable to give the set screw only a one-quarter turn in

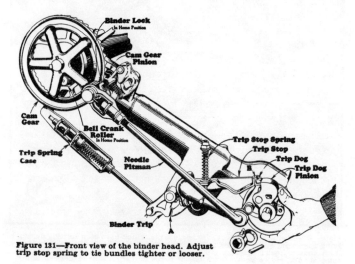

Figure 131—Front view of the binder head. Adjust trip stop spring to tie bundles tighter or looser.

making a test setting. The lock nut must be tightened again after proper setting has been made.

Needle Setting. The needle and the parts that affect its operation should not be tampered with, unless it is obvious that the original settings have been changed or that wear has made an adjustment necessary.

It is the purpose of the needle to place the twine in the twine holder or disk notches (Fig. 133). To do this, when the binder is tripped and turned empty, the needle must press hard against the knife arm. The needle should also set over close to the

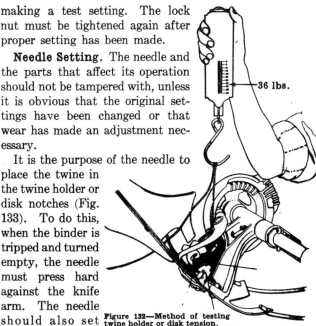

36 lbs.

Figure 132—Method of testing twine holder or disk tension.

knotter pinion when passing through knotter frame, if twine is to be placed in the disk notch properly. If it should be found necessary to advance the needle, this can be done by shortening the needle pitman ("B" in Fig. 131) one or more threads, as the case may require.

Knotter Cam. The knotter hook cam (Fig. 134) presses against the knotter tongue roller and holds the tongue closed when knot is being completed. This is necessary to cause the two ends of the twine to be pulled through to make a complete and secure knot.

Pressure of the cam against the knotter tongue roller is controlled by a coil spring. This spring is often tightened too much. It should be just tight enough to hold the knotter tongue closed when the tying process is being finished.

Knotter Frame Adjustment. (See Fig. 135.) The face of the knotter hook pinion should set up close to the face of the tyer wheel. The wear on these parts is taken up by adjusting eccentrics "A" and "B". When this is necessary, to make the knotter and worm shaft pinions mesh properly with the tyer wheel, care must be taken not to crowd them too tightly into mesh. The knotter frame should be set just close enough to the tyer wheel to permit the face of the knotter hook pinion to rub against the face of the tyer wheel without binding.

Sharpening Twine Knife. For efficient work, the binder operator must keep the twine knife sharp. It needs sharpening frequently, and is brought to the best cutting edge with a carborundum stone (see Fig. 138). A file should never be

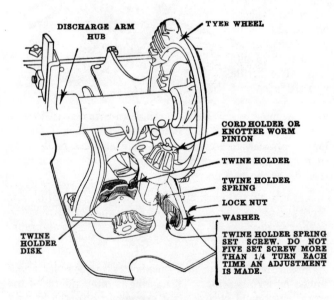

DISCHARGE ARM HUB

TYER WHEEL

CORD HOLDER OR KNOTTER WORM PINION

TWINE HOLDER

TWINE HOLDER SPRING

LOCK NUT

WASHER

TWINE HOLDER DISK

TWINE HOLDER SPRING SET SCREW. DO NOT FIVE SET SCREW MORE THAN 1/4 TURN EACH TIME AN ADJUSTMENT IS MADE.

Figure 133—Close-up view of tying parts.

used, as it produces a rough edge not suitable to cutting twine. The carborundum stone gives a keen edge that cuts easily and accurately.

Hints on Tying Troubles. When starting to locate causes of tying trouble, first examine the twine in the box to see if it is unwinding freely; then follow it through the roller

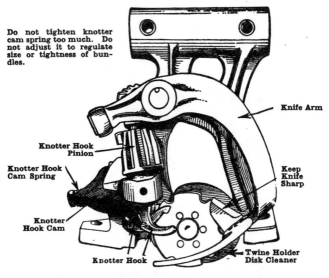

Do not tighten knotter cam spring too much. Do not adjust it to regulate size or tightness of bundles.

Knife Arm

Knotter Hook Pinion

Knotter Hook Cam Spring

Keep Knife Sharp

Knotter Hook Cam

Twine Holder Disk Cleaner

Knotter Hook

Figure 134—Close-up of knotter parts and twine-cutting knife.

tension, twine guides, needle, and disk, watching for conditions that might produce slack or too much tension. Test the twine tension, trip hook tension, and the twine holder or disk tension as directed in preceding paragraphs. Adjust very carefully where necessary.

It is well to remember that uneven, weak, wet, dried-out, or poor twine of any kind will cause missed bundles. Experience has proved that it is poor policy to buy cheap twine of doubtful quality.

Missed bundles may also be due to the condition of the straw. In heavy, tangled, tough straw, which overloads the binder, the needle may carry straws into the twine

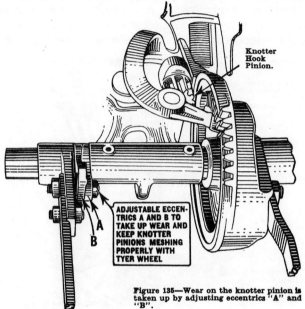

Knotter Hook Pinion.

ADJUSTABLE ECCENTRICS A AND B TO TAKE UP WEAR AND KEEP KNOTTER PINIONS MESHING PROPERLY WITH TYER WHEEL

Figure 135—Wear on the knotter pinion is taken up by adjusting eccentrics "A" and "B".

disk and prevent the twine from entering in the proper manner. A missed bundle results.

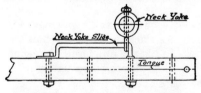

Figure 136—Neckyoke should be at front of the neckyoke slide when the team is moving forward.

This condition can be remedied to some extent by keeping the needle point sharp and the entire needle well polished, setting the grain cover down low, setting the trip hook and trip stop spring to make smaller bundles, and setting the steel breastplate flanges close to the needle.

What to Look for. Some of the usual kinds of defective bands and the cause of each are as follows:

If there is a knot in only one end of the band and the other end is cut square, as though cut off with a sharp knife, the twine holder spring is *too loose.* (See Fig. 133.) This band is usually found with the bundle.

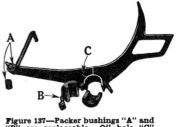

Figure 137—Packer bushings "A" and "B" are replaceable. Oil hole "C" should be filled with oil frequently.

If there is a knot in only one end, and the other end of this band is flattened out, torn, or ragged, the twine holder spring is *too tight.* This band will be found with the bundle.

If there is no knot in either end of the band, the knotter hook cam spring is *too loose.* (See Fig. 134.) This band is usually found with the bundle.

If the band is hanging on the knotter hook and is broken by the bundle being discharged, then the knotter hook cam spring is *too tight.*

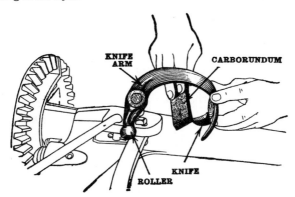

Figure 138—Carborundum stone should be used for sharpening twine knife. Original bevel should be maintained.

If, in picking up a bundle with the band around the bundle, one end slips out, the disk may be slow. This happens on old knotters, due to wear. Put one or two washers just above the worm on the worm shaft to advance the disk just the amount necessary. If this defective knot is made by a new binder, the twine holder spring usually needs to be tightened slightly.

If the band is found tied in slip noose around the bundle with the twine extending from bundle to the eye of the needle, the needle has failed to place twine in the disk. If the roller at needle point is badly worn, loosen and turn half around, and then tighten it.

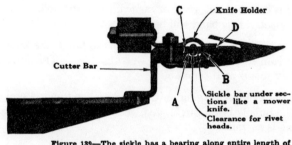

Figure 139—The sickle has a bearing along entire length of the cutter bar at "C", and on all guards at "A". The sickle is also guided by the guards at "B" and holder at "D".

Packers. The packers play an important part in making a compact, neatly shaped bundle. If the bearings become worn, the packers wobble and are inefficient in their work, in which case new bearings should be installed.

Fig. 137 illustrates a binder packer equipped with removable bearings, oil-soaked wood bearing at "B" and tempered steel bearing at "A". This type of bearing makes replacement an easy and inexpensive operation.

Cutting Parts. A cross section of the grain binder cutting parts is shown in Fig. 139. The details of adjustment and repair of the sickle, guards, and knife holders are practically the same as those given for the mower in a following chapter. These details may be studied, if desired, in connection with the study of the grain binder.

Point on Hitching. Binder operators sometimes experience trouble in turning corners with a binder having a tongue truck because short hitching causes tipping of the truck.

This difficulty can be overcome by hitching the team and adjusting the neckyoke slide bolt (Fig. 136) so that the neckyoke works to the front end of the slide when the team is going forward. When turning, the neckyoke can slide back freely on the bolt, and pole team will have more clearance for turning. If the neckyoke does not slide back, the pole team pulls back and down on the tongue when turning, causing the truck to tip. Another advantage of setting the neckyoke slide properly is that it gives the outside horse a chance to travel ahead freely without pulling up on the hitch in turning.

Careful Handling Prolongs Life. A grain binder will give more years of service if it is given more than ordinary care. Thorough oiling, keeping nuts tight, making adjust-

Figure 140—Cutting wheat with a tractor binder.

ments that take up wear, and replacing worn parts before they break are factors that lengthen the life and increase the working efficiency of a grain binder.

As soon as harvest is over and before the binder is stored, it should be cleaned up—all old grease, dirt, and chaff removed, and all cutting and tying parts covered with oil to prevent rust. Working parts should be examined and any found to be worn should be replaced by new parts during slack seasons. Every minute spent in tuning up a binder for the next season may mean a saving of an hour or several hours at harvest time. Fig. 140 shows a grain binder doing a good job of cutting.

Questions

1. *Name the three types of grain binders mentioned.*

2. *What are some of the advantages of a tractor binder over a horse-drawn binder?*

3. *What is of first importance in operating a grain binder?*

4. *How may the tilting lever be used to advantage in "down" grain?*

5. *What is the function of the reel and how should it be run in average conditions?*

6. *What controls the position of the twine on the bundle? Where should the twine be placed to make neat bundles?*

7. *How should the chains and canvases be run for best results?*

8. *Why should the binding attachment be tampered with as little as necessary? Describe the tying process.*

9. *Tell how you would test the twine tension and set it properly.*

10. *What is trip tension and what effect does it have upon the bundles? How is trip tension tested and adjusted?*

11. *What has the twine holder spring to do with efficient tying?*

12. *How and why is the needle advanced?*

13. *Describe the knotter cam and knotter frame adjustments and the purpose of each.*

14. *Name some of the common tying troubles and tell how each is corrected.*

15. *What causes the tongue truck to tip and how may this be remedied?*

Chapter XIV.
COMBINE HARVESTERS

In recent years, the territory in which the combine harvester is used has spread from the wheatfields of the Pacific Coast states to the fields of practically every section of the country. Each year finds this cost-reducing, labor-saving machine proving its value to farmers in new regions. Small, one-man machines for the average farm, and big-capacity machines, with cutting widths as great as twenty feet, make practicable its use in many communities where it has not yet been introduced.

The combine, in many cases, effects a saving of from fifteen to twenty cents per bushel in harvesting costs. It displaces the binder, hand shocking, pitching, and threshing. In one operation, the grain is cut and threshed, the cleaned grain elevated into a storage tank, and the straw scattered on the field to be plowed under for humus, or burned.

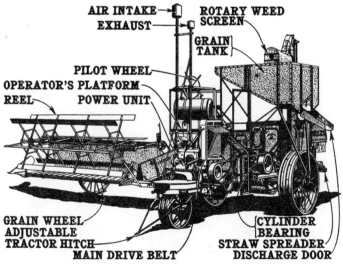

Figure 141—This 12-foot three-wheel-in-line combine will cut and thresh from 25 to 40 acres of grain per day, depending upon the condition and yield of the crop.

The recent advent of the smaller combine with a cutting width of six feet makes practicable the use of this cost-reducing harvesting equipment on small farms where heretofore the use of the combine was not even considered. Larger combines, with cutting widths ranging as high as twenty feet, speed the work and cut harvesting costs on the large farms.

The operator of the combine shown in Fig. 141 controls the machine from his position on the operator's platform. He has a clear view of the grain and stubble and is in position to watch the work his machine is doing. Fig. 142 illustrates the position of the operator in relation to his machine. One man is required to operate the tractor that pulls the combine and one or more men are needed to haul the grain away. When contrasting the size of this crew

Figure 142—Overhead view of a combine, illustrating what the operator sees from his position on the control platform.

with the size of crew required to operate binders and a stationary threshing outfit, it is apparent that a great saving in labor costs can be made with a combine.

The demand for a small combine for the smaller and medium-sized farms where diversified farming is practiced has brought onto the market a one-man combine which is so constructed that it can be used successfully in harvesting many different crops including soy beans in addition to all small grains and many seed crops. A combine of this type is shown in Fig. 143. With this machine, the farmer with an average acreage of small grains can handle his harvest at lowest possible costs.

There are two principal types of combines, the one-shoe and two-shoe machines. The one-shoe machine has but one set of sieves and one fan to clean the grain, while the two-shoe machine has two grain-cleaning units. Because of its lighter weight and greater simplicity, the one-shoe type is gaining in popularity. Fig. 144 shows a cross-sectional view of a one-shoe machine.

Figure 143—The one-man combine harvesting small grain on a Mid-Western farm.

Principle of Operation. The combine performs four major operations—it cuts the grain, threshes or beats the kernels from the heads, separates the kernels from the straw, and cleans the grain, removing dirt and chaff before the grain is elevated into the storage tank.

The cutting unit operates much the same as a binder, with the exception that it is built to cut higher and to deliver the heads and straw into the threshing unit.

The threshing unit, which performs the threshing, separating, and cleaning operations, is similar in construction to the stationary thresher. A cross-sectional view of a one-shoe combine and a description of how the grain progresses through the machine will be found on pages 134 and 135. Careful study of this illustration and the accompanying explanation will provide an understanding of the operating principles of the combine.

The combine may be pulled with either horses or tractor, the latter being the more common and practical method. Most combines have an auxiliary motor which operates the mechanism, leaving the weight of the machine as the only load for the tractor or horses to pull. The weight rolls on four wheels—two main wheels and one front wheel under the separator, and an independent wheel at the outer end of the header platform. This wheel arrangement gives flexibility for work on uneven ground.

Windrowing Method. In many conditions, it is desirable to cut the grain with a windrower and thresh it later with the regular combine equipped with pick-up attachment. Where there are many weeds in the grain, when there is considerable moisture at harvest time, or when the crop ripens unevenly, this method of combine harvesting is used to advantage.

The windrower, shown in Fig. 146, consists of the usual cutting-unit platform, sickle, canvases, and reel. These parts are driven either by a power drive shaft from the tractor or through a ground drive. An opening at the inner end of the platform permits the cut grain to drop out on the stubble in windrow form.

When the grain is properly cured or when the moisture content is sufficiently low, the pick-up unit is attached to the regular combine platform. Its function is to elevate the grain onto the combine platform, and, from this point on, the threshing and cleaning processes are the same as described for the regular combine. The pick-up attachment is illustrated in Fig. 147.

The windrow method of combine harvesting has extended the boundaries within which the combine may be used. Many sections where weeds or rainfall have delayed the introduction of the combine are now using the windrow method with remarkable success.

Operation and Care. All details of the operation, care, and repair of combines will not be given in this text because of the comparatively limited number of students who are called upon to operate machines of this type. Operating directions are furnished by manufacturers with the machines they sell. These directions should be referred to and used by the combine operator.

Certain essentials to successful operation are stressed by all manufacturers, however, and some of these are mentioned here.

The maximum saving of grain and the quality of work done in all conditions depend very largely upon the operator making best use of the adjustments provided for varying conditions. The grain in the tank, the tailings, the straw coming over the straw walkers, and the material going over the shoe reveal the quality of work being done and indicate what adjustments are necessary.

The tractor operator should vary the travel speed to meet conditions. As he approaches a very heavy, down, or tangled condition, he should slow down to give the combine a chance to do a thorough, clean job of separating. In some conditions, it may be advisable to cut less than a full swath, giving

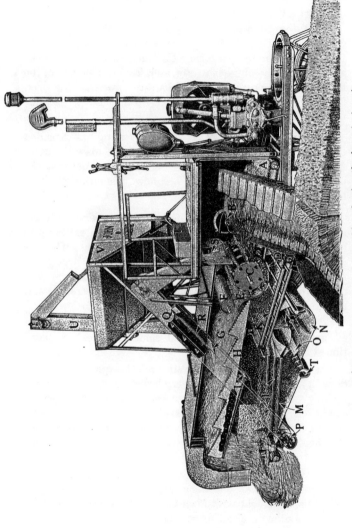

Figure 144—Cross-sectional view of a combine showing the important parts. Explanation of the progress of the grain through the machine appears below.

Cross Section of the Combine

The units are lettered in this cross-sectional view so that the travel of the grain and straw can be followed.

The cut grain is carried by the platform canvas and elevator canvases to the feeder house cross conveyor, "A". This conveyor, with aid of feeder house beater, "B", feeds grain to cylinder, "C".

As the grain travels between cylinder, "C", and concaves, "D", over finger grates, "E", and back against the beater "F", behind cylinder, the greater part of separation takes place. Beater, "F", strips straw from cylinder and deflects grain and distributes straw evenly onto straw walkers, "H". Most of the grain and chaff fall through bottom of concaves, "D", and finger grates, "E", onto elevator, "K" below cylinder.

Straw and remaining loose grain are carried over to straw walkers, "H". Combined metal and canvas retarder, "G", assists heater in retarding straw in rapid movement and keeps grain from being thrown over. Straw is agitated by straw walkers, "H", on its outward movement, and remaining grain falls through opening in walkers and flows back to front of shoe through grain return pans, "I". Straw is then tossed out on spreader, "J".

After the grain and chaff leave elevator, "K", a blast of air from undershot fan, "N", through port, "O", is directed against chaffer, "L", and lower sieve, "M". This, with aid of sieve agitation, blows chaff away and moves tailings to tailings auger, "P". This auger carries them to tailings elevator, "Q", which conveys them to return opening, "R", and back into feeder house.

In tailings elevator, "Q", is sieve, "R", which lets any clean grain through into return spout which delivers it to carrier behind cylinder, preventing cracking of clean grain in going through threshing unit a second time.

Clean grain after dropping through chaffer, "L", and sieve, "M", is carried by clean grain auger, "S", to elevator, "U", on opposite side of machine which delivers it to the grain tank, "V".

the combine every opportunity to do good work. By listening constantly to the sound of the combine motor, the tractor operator can tell approximately how fast he should travel.

The combine operator should not only adjust his machine to hold threshing losses to a minimum, but he should also adjust and operate the platform and reel to reduce cutting losses. He should watch for stones and other foreign material that the platform may gather, and stop his machine before such obstructions reach the cylinder and cause damage.

Safety first should always be the rule when working around a combine. Never attempt to make repairs while the machine is running. Be careful when working around belts and chains. The great majority of accidents around combines result from carelessness.

Proper oiling is of first importance. The large number of bearings in a combine necessitate careful attention to regular and thorough oiling. The combine shown in Fig. 141 is provided with a high-pressure grease gun oiling system which greatly facilitates proper lubrication.

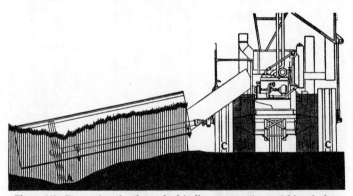

Figure 145—Illustrating the three-wheel-in-line construction and hinged platform of the machine shown in Fig. 141. The grain wheel, "A", follows the contour of the field surface, independently of the position of the separator. Note, also, the position of the tractor wheels, "B", as indicated by the shaded portions just inside of the combine wheels, "C".

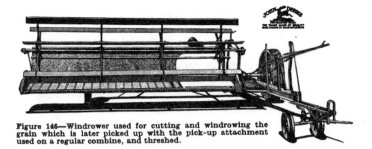

Figure 146—Windrower used for cutting and windrowing the grain which is later picked up with the pick-up attachment used on a regular combine, and threshed.

The upkeep expense on a combine will be greatly reduced if all bolts are kept tight, and belts, canvases, and chains operated at correct tension. Regular inspection of the entire machine will save delays and reduce operating costs to a minimum.

At the end of each season it is important that all dust and chaff be cleaned from the inside and outside of the machine. If not removed, such material will gather moisture and cause steel parts to rust and wood to swell or rot. It will pay the owner of a combine to overhaul and clean the machine thoroughly at the close of each season.

Figure 147—Pick-up attachment for combine attached to the combine platform. The windrowed grain is elevated onto the platform which carries it into the combine in the regular way.

Questions

1. *Are combines used in your community?*

2. *Describe the principle of the combine.*

3. *What are its advantages over other methods of harvesting small grains?*

4. *What are some of the important points to remember in operating a combine?   In storing it?*

5. *Describe the windrow method of combining and the machines used.*

6. *How may the tractor operator aid in doing a clean job of harvesting?*

7. *Why is "Safety First" a good motto for combine operators?*

Figure 148—A hillside combine at work in the far western wheat territory.

Chapter XV.

CORN BINDERS

The corn binder is practically a necessity to the farmer who cuts his corn for ensilage, for fodder, or for shredding. With a corn binder, he can cut and bind from five to seven acres per day while his neighbor is cutting from one to one and one-half acres with a corn knife. The saving in time and labor that can be made with a corn binder pays for its cost in a short time.

Corn binders are built in horse-drawn ground-driven, and tractor-drawn power-driven types. The former type has been used successfully for many years while the latter has been introduced within the last few years. The principal advantages of the tractor binder lie in the fact that it is power driven and does not depend upon traction of a bull

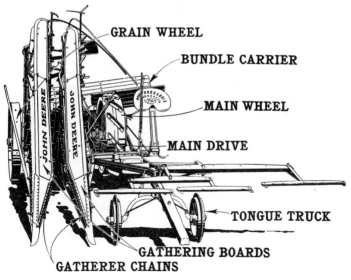

Figure 149—Corn binder with power-driven bundle carrier and tongue truck.

wheel for power. It will operate successfully in practically
any field condition where a tractor will operate. It provides
greater daily cutting capacity because of the faster, steadier
travel of the tractor. Because of their similarity in construction, the operation and care of horse and tractor corn
binders need not be discussed separately.

Like the grain binder, the corn binder gives the best service when it is given better than ordinary care and attention.
It is not difficult to operate and adjust if the operator is
familiar with the more common causes of trouble and knows
how to correct them.

The corn binder is composed of three main units—the cutting, elevating, and binding units. Each has a definite and
vital bearing upon the satisfactory operation of the binder.
Each must be in perfect adjustment if the binder is to do
its best work.

Cutting Unit Important. On the corn binder, the cutting parts are subjected to greater strain than any other part
of the machine. This is due to the size of the stalks, their
comparative hardness, and the fact that the load comes inter-

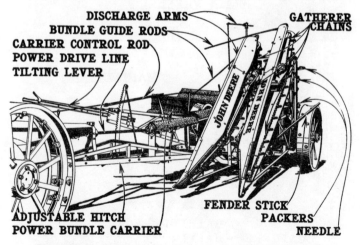

Figure 150—Tractor corn binder with important operating parts named.

mittently as the hills are reached. Even in the best of conditions, there is a much greater strain on the cutting unit of the corn binder than on the cutting parts of a grain binder.

To operate efficiently with the lightest draft under these severe conditions, the cutting parts of a corn binder must be sharp, properly aligned, and set to run smoothly. Fig. 151 shows the complete cutting unit in proper adjustment.

Two stationary knives, one on either side in front of the sickle, aid in cutting the stalks as they approach the sickle. These knives must be kept sharp and set to a shear cut with the sickle. They can be removed and sharpened with very little difficulty. When replaced, the bevel edges should be down. Because of the fact that the side knives and the sickle are often forced to work in the dirt, frequent sharpening of both is necessary. Dull cutting parts increase the draft, add to the strain on the driving mechanism, and may cause clogging of the machine.

The sickle must run freely, yet fit snugly in the guides provided. If the sickle head becomes worn, the knife head guide is adjusted to take up the wear by loosening the two nuts and adjusting the guide in the slotted holes. (See Fig. 151.) When this adjustment is properly made, the sickle and side knives make a shear cut—one of the big essentials to light draft and good work in a corn binder.

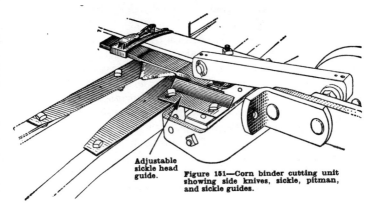

Adjustable
sickle head
guide.

Figure 151—Corn binder cutting unit showing side knives, sickle, pitman, and sickle guides.

Elevating Unit. The elevating unit consists of six carrier chains, two chains on each of the upper gathering boards, and two chains on the lower part of the inner gatherer. The purpose of these chains is to elevate the corn in an upright posi-

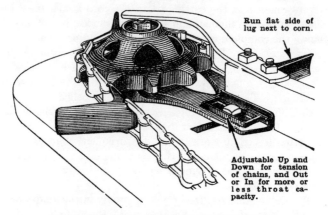

Run flat side of lug next to corn.

Adjustable Up and Down for tension of chains, and Out or In for more or less throat capacity.

Figure 152—Top chain tightener is adjustable two ways.

tion from the sickle to the binding unit. Their efficiency depends upon adjustment to the proper tension and position, which can best be determined by observation in the field.

Convenient tighteners are provided for each of the chains. They should be so adjusted that the chains run freely and are not too tight.

In addition to the adjustment for tightening the top chains, there is an adjustment provided for controlling the throat capacity or the distance between these chains. (See Fig. 152.) In cutting unusually tall corn on a windy day, it is often necessary to set the chains closer together to bring the tops back at the same speed as the butts. If it is desired to retard the tops, the throat capacity is increased by setting the chains farther apart. This is often necessary in cutting short corn.

Another aid to cutting short corn is provided in the small, round retarding spring. It may be set with the end tight against the binder deck.   In this position, it holds the tops back, causing the corn to elevate in an upright position.

The lugs on the elevating chains serve as fingers that carry the stalks along.   Chains that operate opposite to each other should be adjusted so that the lugs alternate rather than match as they move along the throat of the binder.   In this adjustment, they are most efficient, and the danger of ears wedging between lugs and interfering with their work of elevating is eliminated.   Note, in Fig. 152, the lugs in proper adjustment.   The flat side of the lugs should always be run next to the corn, as shown in Fig. 152.

Long steel springs are provided in the lower part of the throat to hold the corn against the lower chains.   They are fastened to adjustable brackets and should be set with just enough tension to hold the stalks into the butt chains.   In weedy conditions, more tension is applied to aid in elevating the extra quantity of material to be handled.

Binding Unit.   The details of the operation and adjustment of the corn binder binding unit are practically the same as those given for the grain binder in a preceding chapter. Adjustments for tying troubles, twine tension, etc., are the same on both.   A review of this text matter will furnish sufficient material for practical study of the corn binding unit.

Field Operation.   The first field adjustment necessary is setting the binder to the height it is desired to cut the corn. In some cases, as in cutting corn infested with the European corn borer, it is desired to cut as close to the ground as possible, while many times high cutting is more practical. Height is controlled by cranks on both the main and grain wheels.   The binder works best when the wheels are set at the same height.   If additional traction is needed, it can be secured by lowering the main wheel.

The grain wheel axle is constructed so that the weight of the binder can be shifted forward or backward to balance the machine properly with any equipment.   (See Fig. 153.) The wheel is shifted to the rear to prevent whipping of the pole, or to place proper weight on the tongue truck. It is shifted forward to relieve neck weight when the tongue truck is not used.

The binder is tilted with the tilting lever to adjust the position of the gatherers with relation to the ground. This setting is governed by the condition of the corn to be cut. If the corn is down, the gatherer points should be run close to the ground.

The butt pan, upon which the butts slide from the sickle to the binding head, is adjusted up or down at the rear with the pan lever. The binder should be operated with the pan as low as possible, raising it only when it is necessary to place the band closer to the butts.

Bundle Carrier. The power bundle carrier is set into operation by tripping a foot lever. It delivers the bundles beyond the path of the horses as they make the next round. This eliminates the waste caused by the horses tramping the corn and does away with the hard work of dumping the bundles and returning the carrier, which is necessary when the old-fashioned bundle carrier is used. A safety clutch in the carrier drive removes the possibility of breakage should the forward motion of the carrier be checked for any reason. The spring tension on the clutch is adjustable to meet varying loads that may be carried.

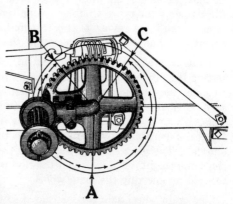

Figure 153—Grain wheel adjustment for balancing corn binder, showing grain wheel set to rear between points "A" and "B" to keep stiff pole from whipping or to keep enough weight on tongue truck. Set wheel to front between "A" and "C" on binder with stiff pole, but without carrier.

Bundle Elevator. Some manufacturers of corn binders furnish bundle elevators as extra equipment for their machines. The purpose of this attachment is to elevate bundles direct from the binder onto the wagon which is driven along beside the outfit. A tractor binder with bundle elevator is shown at work in Fig. 154, while a detail view of this equipment is illustrated in Fig. 155.

The bundle elevator is used principally for handling corn that is to be put into silos. The usual practice where this elevator is not used is to drop the bundles on the ground with the regular bundle carrier, then lift them onto the wagons by hand and haul to the ensilage cutter. Lifting tall, green bundles is hard, tiresome work. The bundle elevator does away with this job.

The bundles fall directly onto the elevating chain of the bundle elevator as they are discharged from the binding unit. They are elevated onto the wagon where one man can place them on the load. Handling bundles in this manner is easy compared with lifting them onto the wagon. And, too, the loose leaves and ears are elevated with the bundles—all the crop is delivered clean and dirt free.

Figure 154—The bundle elevator on the corn binder is a great labor-saver. It elevates bundles, loose ears, and leaves directly onto the wagon.

Oiling Important. Frequent and thorough oiling of all bearings, chains, and working parts of the corn binder will reduce wear and add to its efficiency. The operator must be sure to keep the bearings on the binding unit oiled properly.

Before the corn binder is stored, it should be cleaned of all dirt and oil accumulations and inspected for worn parts. If new parts are needed, they should be ordered and attached during slack seasons, ready for the next year's work.

Figure 155—Overhead view of a tractor corn binder with bundle elevator and wagon hitch. For filling silos, this type of outfit cuts man labor, saves time, and saves loose ears and leaves.

Questions

1. *What are the main advantages of a tractor corn binder over a horse-drawn binder?*

2. *Name the main units of a corn binder and tell the function of each.*

3. *Why is it important that cutting parts be kept sharp and in proper adjustment?*

4. *Why are the elevating chains an important factor in good binding?*

5. *What adjustments are necessary in cutting short corn? In cutting tall corn on a windy day?*

6. *How is height of cutting controlled? How is the binder balanced?*

7. *What is the purpose of the tilting lever?*

8. *What are the advantages of a power bundle carrier?*

9. *Tell the more important points in properly caring for a corn binder.*

Chapter XVI.
ENSILAGE HARVESTERS

We have seen how the corn binder, especially when equipped with bundle elevator, materially reduced the amount of heavy work necessary in cutting corn for ensilage. However, with this equipment, the work of unloading heavy green bundles and feeding them into the ensilage cutter still remains as a tedious part of putting up the crop.

Where corn is to be cut for ensilage, the field ensilage harvester solves the problem with the minimum of work, at extremely low cost, and enables the silo owner to harvest the crop at just the right time to make high-grade ensilage.

In principle, the field ensilage harvester embodies two separate units. The power-driven harvesting unit consists of the gathering and cutting parts, somewhat similar in design to corresponding parts used on the corn binder, in combination with a high-speed cutting unit designed to

Figure 156—From standing corn to top-grade ensilage in one operation.

cut the corn into lengths suitable for ensilage. A second unit, the blower, is stationed at the silo; its function is to blow the ensilage into the silo.

In operation, the gathering and elevating parts of the harvester unit cut the corn and deliver it to the feed table, where the feed conveyor carries it into the cutting unit. Here, revolving knives, driven by the tractor engine through the power shaft, cut the corn into suitable lengths, and pass the freshly cut ensilage into the wagon elevator which carries it up and drops it into the wagon, driven along beside the outfit.

The blower unit, stationed at the silo, finishes the job by elevating the ensilage into the silo.

The details of operating and adjusting the gathering, cutting, and elevator parts of the field harvester are so

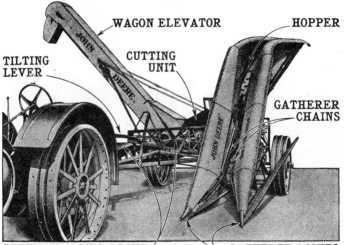

Figure 157—The harvesting unit of the ensilage harvester with important parts indicated.

similar to those given in a preceding chapter for comparable parts of the corn binder that a further discussion of them is not necessary. Most important in caring for the cutting unit, which runs at high speed, is providing sufficient lubrication to all parts and keeping all bearings tight. In most present-day harvesters, the knives are of the self-sharpening type which require but infrequent attention, other than to maintain the stationary blade in proper relation to the revolving knife head.

The blower, like the cutting unit of the harvester, should be kept thoroughly lubricated to insure efficient operation and long life. As is the case with all high-speed equipment, especial attention should be given to proper adjustment and lubrication of all parts, and to keeping all nuts tight.

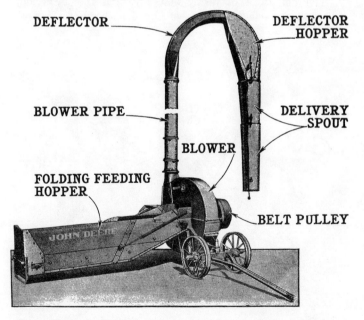

Figure 158—The blower unit is stationed at the silo to handle the ensilage from wagon to silo.

When the season's work is done, both units should be cleaned thoroughly and inspected for worn parts which should be replaced with new parts during the slack season.

Questions

1. *Review the paragraphs on the cutting and elevating units of the corn binder as a key to servicing similar parts on the harvesting unit of the field ensilage harvester.*

2. *What are the advantages of using the field ensilage harvester?*

3. *Describe the main parts of the harvester unit and mention their purpose.*

4. *How is the cutting unit driven?*

5. *What is the function of the blower?*

Chapter XVII.
CORN PICKERS

Wherever corn is husked, either for the market or for feeding purposes, the corn picker is more generally used each year. The hard work of husking by hand is fast being replaced by the easier, faster, and less expensive mechanical method.

The corn picker has been refined and simplified, mechanically, to a point where its operation is not difficult. With the aid of the manufacturer's instruction book, furnished with each new machine, the average farmer finds the corn picker comparatively easy to operate.

When mechanical corn pickers were first introduced, the one-row, horse-drawn, ground-driven type was the only type manufactured. With the advent of tractors and power farming came the power-driven one-row, and later, the power-driven two-row pickers which have greatly increased the corn

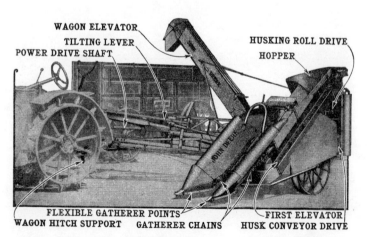

Figure 159—Corn picker with the more important parts indicated and named.

picking capacity of one man.   With a power-driven picker equipped with wagon hitch, one man has control of tractor, picker, and wagon.

A comparatively recent innovation is the push-type or mounted-type two-row picker, such as illustrated in Fig. 160, which, when attached to the tractor, makes up a compact, easily handled one-man picking outfit.   Its advantages lie in the fact that no hand-picking is necessary in opening fields, that the corn is handled in a direct line from the snapping rolls to the wagon elevator, and in the fact that, with gatherers in front of the tractor, the operator has a good view of the work at all times.   With its rear delivery, a direct center hitch is provided for the wagon, thereby eliminating side draft.   In addition, the mounted picker is easily transported from field to field, since its control is just as easy as driving the tractor.

How They Work.   The function of corn pickers is to snap the ears from stalks, remove husks and silks, and deliver the cleaned ears into a wagon.   To do this, three main units are required—the snapping, husking, and elevating units.   Ef-

Figure 160—Opening a field with a push-type two-row picker.

ficiency of the picker depends to a great extent upon the correct adjustment of the snapping and husking units.

Power to operate the corn picker is furnished direct from the tractor engine through a power drive shaft. Thus the tractor engine, running at steady speed, insures uniform power for operating snapping, husking, and elevating units.

Figure 161—Overhead view of a two-row corn picker with tractor and wagon attached. The lines indicate the position of the rows in relation to picker, tractor, and wagon.

The principle of operation of one-row and two-row pickers is so similar that both will be covered in the following text.

Snapping Unit. As the cornstalks advance between the gatherers, they are drawn into the snapping rolls with the aid of gathering chains, one on the inner and two on the outer gatherer. The stalks pass between the rolls while the ears are snapped off and carried to the husking unit. Snapping rolls of a one-row picker are illustrated in Fig. 162.

The snapping rolls are adjusted according to the condition of the corn; if damp, rolls are run close together, but not touching; if dry, the rolls are set to run farther apart. They should never be run closer together than necessary for good work—to do so increases the draft. The desired setting is secured by adjusting the notched bracket at the lower end of the outer roll.

A tension spring at the upper end of the rolls permits obstructions to pass between the rolls without breakage of parts. It should be set with just enough tension to keep the gears well in mesh. Too much tension may cause breakage.

Husking Rolls. The husking rolls operate in pairs. They are held together under spring pressure at either end and are adjusted by tightening or loosening two nuts. There should be just enough tension on the rolls to cause them to grasp the husks when the smooth surfaces come together. Too much tension will unnecessarily increase the draft.

One of the husking rolls in each pair is provided with husking pegs—twenty in number—which loosen the husks; the other roll has a recess for each peg in the companion roll, providing clearance as they revolve. The pegs are easily replaced when worn or broken.

Movement of the ears over the rolls of the machines illustrated in this chapter is controlled by a revolving conveyor retarder (see Fig. 164). The ears are held down to give the husking rolls ample opportunity to remove the husks.

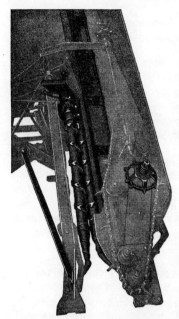

Figure 162—Above is shown a detailed drawing of the snapping unit and first elevator of a one-row picker. The galvanized top has been removed from the gatherer to show the working parts.

Wagon Elevator. The elevating unit carries the husked ears from the elevator hopper into the wagon. The hopper, into which the ears fall after the husking process is finished, is large enough to hold the surplus corn delivered to it when the elevator is stopped to change wagons and when turning at the ends.

The elevator drive is provided with a safety clutch which prevents breakage of chains and other parts if the elevator becomes clogged. The same type of clutch is also provided on the gatherers, first elevator, husking rolls, and husk conveyor for the same purpose. The slip clutch that protects the main drive is shown in Fig. 163. The springs controlling these clutches should have just enough tension to hold the clutch in contact when the parts are working under the normal load. If clogging occurs, and the added load throws the clutch out, the operator should stop immediately, locate and correct the trouble.

Oiling Important. The corn pickers shown at work in Figs. 160 and 165 are provided with roller and ball bearings at the important wearing points, all of which are supplied with facilities

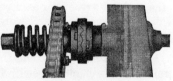

Figure 163—Slip clutches protect the main working parts of the corn picker against breakage, should congestion occur.

that make oiling easy.   Thorough oiling of these bearings and all other moving parts is highly important to light draft and long wear.

All chains should be brushed with a light oil occasionally and should be run just tight enough to do their jumping.

That extra care in handling will prolong the life and increase the efficiency of the corn picker should be the first lesson of every prospective operator.

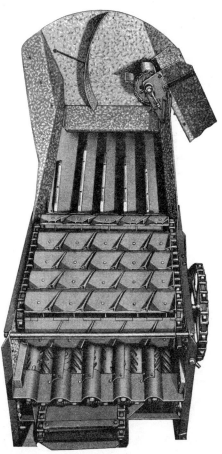

Figure 164—The husking rolls of a two-row corn picker showing the 5 sets of retarding plates that hold the ears against the husking rolls, aiding in removing the husks.

Questions

1. *Describe two types of corn pickers and outline their differences.*

2. *Name the advantages of a corn picker over the hand-picking method.*

3. *Describe the operation of a corn picker.*

4. *What are the three main units of a picker and what is the purpose of each?*

5. *Why are safety clutches used and how do they operate?*

6. *Tell the important points in caring for a corn picker.*

Figure 165—Two-row picker at work in an Illinois field.

Chapter XVIII.
POTATO DIGGERS

Modern potato diggers have taken much of the drudgery out of the potato harvest. They have reduced the waste common to the use of ordinary plows or hand-digging methods. A digger is practically a necessity to the economical production of potatoes for the markets.

There are two types of diggers in common use, the horse-drawn elevator type, shown in Fig. 166, and the tractor-drawn, tractor-powered elevator type, shown in Fig. 167. These types are furnished with various equipment units which adapt them to practically all harvesting conditions.

Principle of Operation. The wide steel shovel, set to run at a safe depth below the potatoes, raises the potatoes, dirt, vines and all, onto the elevator. Depth is regulated by adjusting the lifting lever, and by adjustment of beams.

The main object of the potato digger is to deliver clean potatoes on top of the soil. As stated above, the shovel, digging under the potatoes, loosens the soil and, by the forward motion of the digger, the soil, potatoes, and vines are

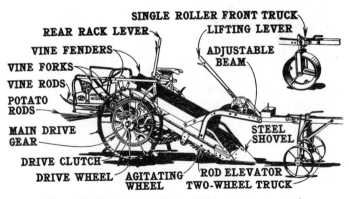

Figure 166—Elevator potato digger with important parts named.

carried onto the elevator. The elevator is a continuous chain moving toward the rear of the machine. The earth drops through the elevator, aided by the agitating or up-and-down movement of the elevator. This agitation can be increased or decreased according to soil conditions by interchanging the smooth rollers and oblong agitating sprockets provided.

The potatoes and vines are carried to the rear of the machine where they are deposited on top of the soil. There are three kinds of rear racks for diggers. The extension elevator is constructed similarly to the main elevator except that it is operated separately. The continuous elevator, as its name implies, is a continuation of the main elevator, extending to the rear. The agitating rear racks, shown on the digger illustrated in Fig. 166, separate the vines from the potatoes, placing each in a separate row. Having the potatoes in a narrow row, separated from the vines, is an advantage in picking them up.

Tractor-Drive Diggers. The tractor-drive digger (in both one- and two-row sizes) is similar in principle to the horse-drawn digger, shown in Fig. 166, with the exception

Figure 167—Two-row tractor-drive digger.

of all parts that transmit power through the ground wheels. In place of these parts, a shielded drive shaft, protected by a slip clutch, is used to transmit power direct from the tractor engine to operate elevators and separating mechanism.

The tractor-drive digger is especially advantageous under difficult conditions, as the speed of its elevator is not affected by wheel slippage or by sudden slowing of forward travel.

Questions

1. *Name two types of potato diggers and tell the advantages of each.*
2. *Which type is most popular in your community?*
3. *Describe the action of an elevator digger.*

Chapter XIX.
MOWERS

Mowers are in use in all sections of the country, and their operation, care, and repair should be a matter of general knowledge among farmers. Heavy draft, ragged cutting, and excessive breakage can often be avoided by using the maximum of care in the oiling, adjusting, and replacing of parts. A smooth-running, clean-cutting mower gives real satisfaction to the operator and requires less power from horses or tractor. Fig. 168 shows a popular style of enclosed-gear mower with all main parts named.

The cutter bar and its parts, including the pitman shaft, (see Figs. 169 and 170) make up the most vital unit in mower operation. These parts do all the work of cutting; draft, repair costs, and length of life of the mower depend upon the proper setting and care given them.

Register of the Knife. The knife is the heart of the mower. Its sections must be sharp and firmly riveted to the knife back; the guards, wearing plates, and knife holders must

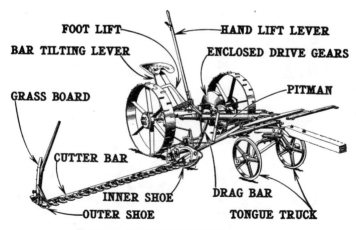

Figure 168—A popular type of enclosed-gear mower.

fit to it perfectly, holding it to a shear cut with the guard plates, if its work is to be efficient. Knife head guides must be properly set and bolted tight.

Register of the knife refers to the position of its sections in relation to the guards when the knife is at the outer end of its stroke and at the inner end of its stroke. The sections should be in the center of the guards when at the extremes of the strokes.   Fig. 170 shows the knife in register on the inner end of the stroke.

If the knife does not register on its outward stroke, that is, if the sections do not reach center of guards, part of the vegetation is not cut.   The results are an uneven job of cutting, an uneven load on the entire mower, heavier draft, and often, clogging of the mower knife.   An incomplete inward stroke will result in the same troubles.

To test and correct the register of a mower knife, raise the tongue to working position—32 inches from underside of front end of tongue to the ground —and turn the flywheel over until knife is at outer end of stroke. It is necessary

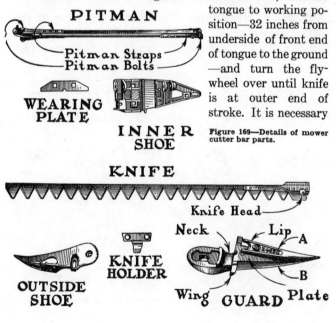

PITMAN

Pitman Straps
Pitman Bolts

WEARING PLATE

INNER SHOE

Figure 169—Details of mower cutter bar parts.

KNIFE

Knife Head

Neck   Lip   A

OUTSIDE SHOE

KNIFE HOLDER

B

Wing   GUARD Plate

that pitman straps at both ends of pitman are tightened properly before making the test. If the sections do not center, an adjustment should be made. On most makes of mowers, register is obtained by adjusting the brace bar at the flywheel bowl. One complete turn of the brace bar will make 1/8-inch difference in register. In addition to this adjustment, washers are provided at both ends of the drag bar bearing on the mower shown in Fig. 168, so that knife will be properly centered without destroying the correct lead in the cutter bar. By transferring more or fewer washers from one end of the yoke to the other as may be necessary, and at the same time adjusting the brace bar, proper setting is obtained.

Cutter Bar Alignment. All new mowers have a certain amount of lead in the cutter bar; that is, the outer end is ahead of the inner end to offset the backward strain produced by the pressure of cutting and to permit the knife and pitman to run

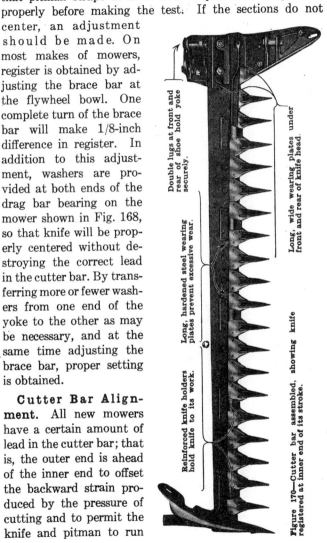

Figure 170—Cutter bar assembled, showing knife registered at inner end of its stroke.

in a straight line. As the mower wears and parts become loose, the outer end of the bar lags back until the knife is running on a backward angle, causing undue wear and breakage of cutting parts. The outer end of the cutter bar should be ahead of the inner end 1 to 1-1/4 inches on 4-1/2-foot mowers; 1-1/4 to 1-1/2 inches on 5-foot mowers; 1-1/2 to 1-3/4 inches on 6-foot mowers; and 1-3/4 to 2 inches on 7-foot mowers.

To determine the lead or lack of lead in a cutter bar, raise the end of the tongue (underside) 32 inches from the ground. Then tie a cord to the oil cap on the pitman box ("A" in Fig. 171). Stretch the cord over the center of the knife head, as shown at "B", Fig. 171; the amount of lead in the bar can be determined at point "C". The upper illustration in Fig. 171 shows a 5-foot bar with the proper lead—approximately one inch.

The lower illustration in Fig. 171 shows a bar with nearly 4 inches of lag—a condition that would result in heavy draft and poor work were it not corrected. With the cord tied to the oil cap on the pitman box, "D", and stretched over the center of knife head at "E", the position of the outer

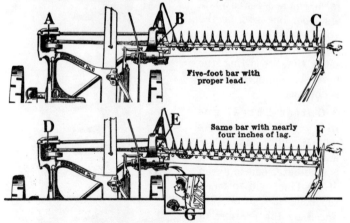

Figure 171—Overhead view of a mower showing how to determine lead or lag in the cutter bar.

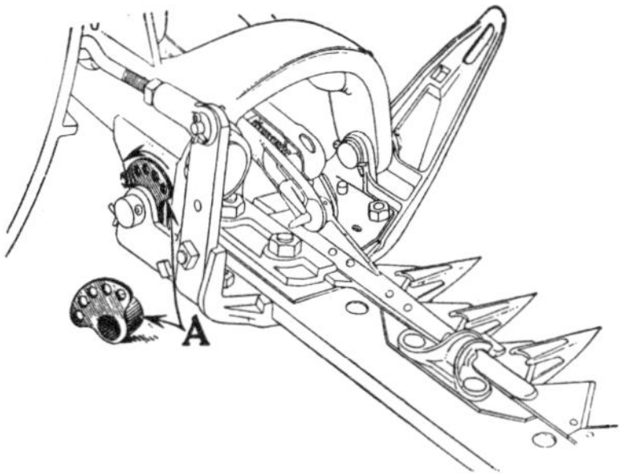

end of the bar at "F" is seen to be 3 inches behind the straight line, or approximately 4 inches behind the position at which it should be maintained for the correct amount of lead.

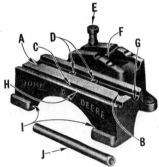

Figure 172—Eccentric "A" is adjusted to left to take up lag in cutter bar.

Lag in the cutter bar of the mower shown in Fig. 168 is removed and the bar brought up to proper position by turning an eccentric bushing ("A" in Fig. 172) to the left until proper alignment is obtained.

After hard usage, enough wear may occur to create excessive free motion of cutter bar. In such cases, new parts may be necessary for effective adjustment.

Pitman Adjustment. The mower operator must keep the bolts that connect the pitman to the pitman box and knife head at proper tension for good work. If pitman bolts are too tight, particularly the knife head bolt, the draft will be increased. The knife head must have a free ball-and-socket action in the pitman straps (see Fig. 169) to accommodate tilting of the bar and the up-and-down movement of the inner and outer ends of the bar in going over uneven ground. If

Figure 173—A convenient block for removing and replacing guard plates, knife sections, flywheel wrist pins, and for straightening knives. "A" and "B" are holes used in riveting wrist pin to flywheel. "C" indicates a hole through which sheared rivets are driven. Removable, hardened riveting posts are shown at "D". "E" is the guard plate riveting post. "F" indicates hole through which old rivet is driven. "G" is a groove for the knife back. When shearing sections, the knife back rests on edge "H". The grooves "I" steady the knife. "J" is the rivet set used in completing the job of riveting.

the bar is tilted low, a tight pitman connection tends to hold the knife sections away from guard plates. This causes excessive wear and allows the grass to get between sections and guard plates.

If the pitman bolts are permitted to become too loose, or the strap rivets loosen, the pitman and knife are sub-

jected to excessive vibration, which results in heating of pitman box, breakage of parts, and abnormal wear. Operating

Fig. 174—The efficiency of the mower is seriously impaired by imperfect guard plates.

the pitman with the hand will usually show whether or not it is in proper adjustment.

Adjustment and Repair of Cutter Bar Parts. (See Figs. 169 and 170.) When the pitman is properly adjusted and all cutter bar parts are set as they should be, the front end of every knife section rests smoothly on the guard plate, in position to make a shear cut. To maintain this ideal condition, guards and guard plates, wearing plates, and knife

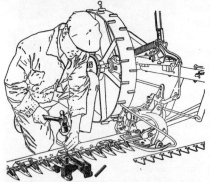

holders must be in good condition and correctly set. If these parts become loose or badly worn, the knife will flop around in the cutter bar, chewing and tearing the grass instead of cutting it, causing the mower to pull hard and in-creasing the possi-bilities of breakage.

Fig. 175—Replacing guard plates in the field with the repair anvil, shown in Fig. 173.

The guard or ledger plates have a very important function in the cutting action of the mower. They act as one-half of the shear, the knife sections acting as the other half. If sections and plates are not sharp or do not fit closely together, the result is similar to that produced by a dull or loose shears in cutting cloth. Guard plates should be replaced when broken or worn dull (Fig. 174) and the guards aligned to give a shear cut on every plate. Figs. 173 and 175 show a convenient block for replacing plates with guards either on or off the cutter bar.

A dull or improperly ground knife reduces the efficiency of the mower, results in ragged cutting, excessive and unnatural wear, and extremely heavy draft. By actual dynamometer tests, a dull or improperly ground knife may increase draft of the mower as much as 30 per cent over the normal draft of a new or properly ground knife.

The angle at which the sections work with the guard plates and the angle of the cutting bevel on new sections have been worked out by years of

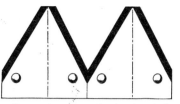

New sections—proper bevel and angle for good work.

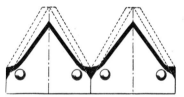

Sections properly ground. Even after repeated grinding, proper bevel and angle are retained.

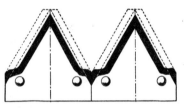

Improperly ground sections; narrow bevel and wrong angle which changes the angle of "shear."

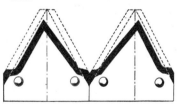

Sections ground off center, destroying the register of blade in guard.

Fig. 176—The right and wrong ways to grind mower knives. Dotted lines shows outline of new sections.

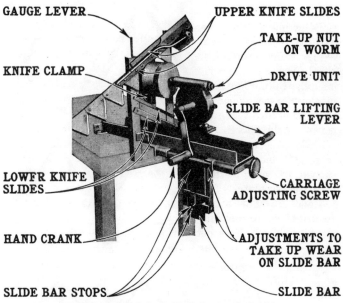

GAUGE LEVER

UPPER KNIFE SLIDES

TAKE-UP NUT ON WORM

KNIFE CLAMP

DRIVE UNIT

SLIDE BAR LIFTING LEVER

LOWER KNIFE SLIDES

CARRIAGE ADJUSTING SCREW

HAND CRANK

ADJUSTMENTS TO TAKE UP WEAR ON SLIDE BAR

SLIDE BAR STOPS

SLIDE BAR

Fig. 177—Knife grinder and knife in upper knife slides.

trial and experience; they are practically standard on all mowers. When grinding the knife, it is of utmost importance that these angles be retained if the knife is to be restored to its full efficiency. The angle at which the section meets the guard plate must be such that the grass will not have a tendency to slip away. The bevel of the

Figure 178—When removing and replacing knife sections, a solid base must be provided to prevent bending or breaking knife back.

section is highly important, as an abrupt edge will tend to dull easily and chew the grass, thereby increasing draft; too wide a bevel will cause the section to nick easily. (See Figure 176.)

The type of knife grinder shown in Fig. 177, because of its better work, greater speed, and easier, more simple operation, is fast replacing the grindstone method of reconditioning knives. With this type of grinder, the operator simply slips the knife into slides, locates the proper setting with the gauge, clamps the knife into place, and brings the revolving stone into contact with the section by turning the horizontal adjustment screw. Up and down movement of the stone is controlled by the vertical adjustment lever. Most manufacturers provide foot-, motor-, and engine-power grinders in addition to the hand-power grinder shown.

When knife sections have been ground to the point where the efficiency of the mower is impaired, or when sections have

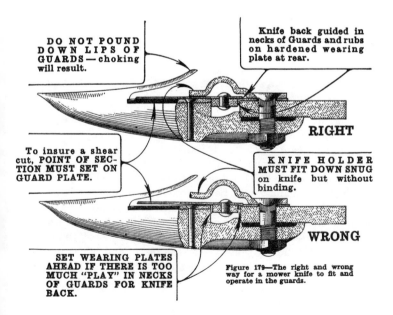

DO NOT POUND DOWN LIPS OF GUARDS — choking will result.

Knife back guided in necks of Guards and rubs on hardened wearing plate at rear.

RIGHT

To insure a shear cut, POINT OF SECTION MUST SET ON GUARD PLATE.

KNIFE HOLDER MUST FIT DOWN SNUG on knife but without binding.

WRONG

SET WEARING PLATES AHEAD IF THERE IS TOO MUCH "PLAY" IN NECKS OF GUARDS FOR KNIFE BACK.

Figure 179—The right and wrong way for a mower knife to fit and operate in the guards.

Figure 180—A timesaving hook-up of power mower with tractor drawn mower, cutting from 50 to 70 acres per day.

been broken, the worn or broken plates should be sheared off, as shown in Fig. 178. With knife back resting solidly on the block, strike back edge of section a sharp blow. This operation will shear the rivets without damaging knife back. Driving out the rivets with a punch not only enlarges the holes, but weakens the knife back.

In replacing the sections, be sure the rivets are *tight* and properly rounded.

Aligning the guards is an important and exacting operation. A new knife, or a straight one that is not badly worn, should be used in testing and setting the guards. Insert the knife and set each guard up or down, as necessary, to make a shear cut between knife section and guard plate. Guards are malleable iron and can be bent without breaking by striking at the thick part, just ahead of plate when guard bolt is tight. Guard wings should also be aligned, making a smooth surface for knife back to work against. Position of guard points should not be considered—the plates and wings are the important units that must be aligned. See Fig. 179 for complete information on proper alignment of guards to produce a shear cut.

It is advisable to replace badly worn wearing plates (see Figs. 169, 174, and 179) when guards are repaired. The wearing plates hold the sections in correct cutting position, but when worn, they permit the sections to rise at front end, causing clogging and ragged cutting.

The knife holders hold the sections down against the guard plates. They must be set close enough to the sections to hold them firmly in position when cutting, yet not tight enough to cause binding and heavy draft.

When necessary to set the holders down, the knife should be pulled out—holders should never be set down with knife under holder. Starting at holder next to the outer shoe, set each holder down with a hammer, tapping it lightly.

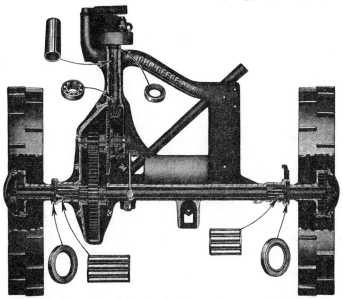

Figure 181—Cross-sectional view showing oiling system, gears, clutch, and bearings of the mower shown in Fig. 168. Color indicates parts that are automatically oiled.

Then move the knife under holder to test the adjustment. If it tends to bind, leave knife under the holder and hit the holder on the flat surface between the two bolts. Proper setting of each holder must be made before moving to the next one.

Lifting Spring. There should be enough tension on the lifting spring to cause the bar to rise easily and move steadily over the ground. With too much tension, the bar will not follow uneven ground, and the inner end may be held up after it has passed over a mound or other obstruction. When properly adjusted, the lifting spring carries the bulk of weight of the cutter bar on the wheels, increasing the traction and reducing friction between bar and the ground.

Clutch Adjustment. When clutch parts become worn, it is often necessary to make minor adjustments for good work. The clutch shifter rod is adjustable to take up wear and keep the clutch engaged full depth in drive gear. If the mower does not go out or stay out of gear when shifter lever is moved down, it is evident the clutch shifter rod adjustment must be shortened. This is done by loosening the lock bolt on the clutch pedal and turning the clutch throw-out sleeve to the left. After proper adjustment is made, tighten lock bolt and secure with cotter key.

When the clutch lever is up and the clutch meshes full depth with gear, the clutch shifter yoke should be free in the clutch and not bind against either side of the groove in the clutch.

Operation and Care. Mowers require a considerable amount of attention and care when at work. Following are a few hints for mower operators:

See that all moving parts work freely before putting the machine in the field. Keep all nuts tight.

Use plenty of good grade oil, and never let wearing surfaces become dry.

Oiling of the mower shown in Fig. 168 is greatly simplified because all gears are enclosed and running in oil. In addition, the axle, wheel, gear, countershaft, and pitman shaft bearings are oiled automatically from the gear case, as shown in Fig. 181.

In dry, dusty, or sandy conditions, the cutting parts usually work best without oil.

The mower is in correct working position when underside of the tongue at the front end is 32 inches from the ground.

It is advisable to keep the horses close together by shortening the inside lines.

In summary, it can be said that the repairing of mower cutter bars is generally put off too long and that greater care in oiling and adjusting the important wearing parts will add to the length of the mower's life.

Questions

1. *What is meant by register of a mower knife? What effect does lack of register have upon the work of a mower?*

2. *Tell how you would register a mower knife found to be out of register.*

3. *What is meant by "cutter bar alignment"? How would you test for alignment and how would you take up "lag," if present?*

4. *Why is it necessary to keep the pitman straps in proper adjustment?*

5. *Name the important parts of the cutter bar.*

6. *What is a "shear cut"? What parts must be in proper adjustment for this ideal cutting condition?*

7. *Describe proper procedure for grinding mower knives.*

8. *How are the guards aligned?*

9. *What is the function of the knife holders?*

10. *How would you set the lifting spring for good work?*

11. *What are the most important points in properly caring for a mower?*

Tractor Mowers

Because of the fact that the general purpose type of tractor must do all farm jobs if it is to displace horses on the farm, manufacturers now provide power-driven mowers that are quickly attached to and detached from the tractor. Greater mowing capacity is possible with a power-driven machine because of the steadier rate of travel and the wide cut of the cutter bar. The machine shown in Fig. 182 has a seven-foot cutter bar—a width that provides maximum cutting capacity.

Operation. It is little more difficult to operate a tractor mower than it is to operate a horse-drawn mower. The cutting parts are practically the same in construction, and require the same care and adjustments for maximum efficiency. The knife must register, the cutter bar must be in

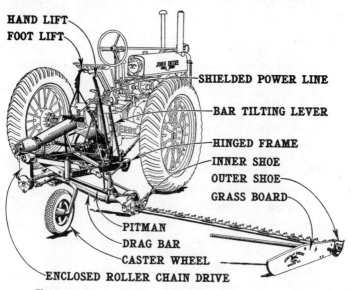

HAND LIFT
FOOT LIFT
SHIELDED POWER LINE
BAR TILTING LEVER
HINGED FRAME
INNER SHOE
OUTER SHOE
GRASS BOARD
PITMAN
DRAG BAR
CASTER WHEEL
ENCLOSED ROLLER CHAIN DRIVE

Figure 182—Tractor mower, trailer type, attached to general purpose type of tractor.

alignment, and there must be a shear cut of sickle and guard plates if good work is to result. The means of obtaining proper adjustment of these parts are discussed on preceding pages.

Power for operating the sickle is transmitted from the power take-off to the pitman through a drive shaft and enclosed roller chain drive. A slip clutch in the power line releases automatically when the sickle clogs or when the strain of cutting in tough material is too great. Breakage of parts due to overstrains is practically eliminated if the operator keeps the tension on the slip clutch properly adjusted. It

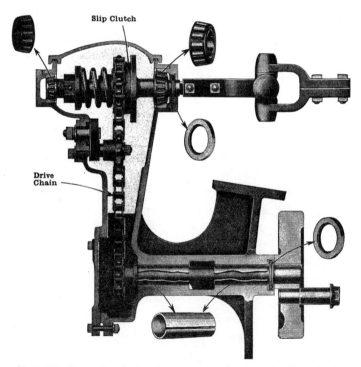

Figure 183—Cross-sectional view of tractor mower showing main drive, clutch, and bearings. Color indicates oiling of enclosed mechanism.

should be set tight enough to do ordinary work without slipping, but loose enough to slip easily if clogging occurs. Extreme care should be exercised in making this adjustment—directions furnished by the manufacturer should be followed closely.

When the cutter bar of the tractor mower, shown in Fig. 182, hits an obstruction, a spring release unlatches and permits the bar to swing back, protecting the mower against breakage. The spring release holds the bar in cutting position under all normal conditions but releases when an unusual pressure or shock is encountered. The amount of pressure required to cause the lock to release is adjusted by tightening or loosening a spring tension. The operator must be careful to avoid getting this spring too tight, as such a condition may cause breakage should the lock fail to release.

After the bar swings back, it may be returned to operating position by backing the tractor.

For ordinary field operation, the foot lift raises the bar high enough to meet field conditions. If the bar must be raised higher, the hand lever may be used. The tilting lever controls the tilt of the bar in relation to the ground. In some conditions, it is necessary to raise the guard points higher than in others. The tilting lever provides this adjustment.

Thorough oiling and careful attention to proper adjustment and repair will increase the efficiency and lengthen the life of a tractor mower.

Questions

1. *How is the tractor mower driven?*

2. *What is the purpose of the slip clutch in the power line? Why is it necessary that it be properly adjusted?*

3. *What happens when the cutter bar hits an obstruction?*

4. *How is the bar raised? Tilted?*

Chapter XX.
HAY RAKES, LOADERS, AND PRESSES

Hay is a highly perishable crop. It must be cut at the right time, cured properly, and handled carefully from field to feeding rack if its maximum feeding value is to be retained. This is especially true of legume hay, such as alfalfa, clover, and soy bean. Timothy, blue grass, and wild hay are less perishable and do not require such exacting methods of handling.

The greater part of the feeding value of legume hay is contained in the leaves. Every operation in curing and handling must have as its main purpose conservation of the leaves, along with thorough curing of the stems. High-grade legume hay should have its natural green color, be fresh and sweet, and should retain all of the leaves on the stems without shattering when the hay is handled.

Modern machines and modern methods of handling hay are used to great advantage in increasing the feeding and market value and decreasing the cost of producing legume hay. The side-delivery rake has gained in popularity in recent years,

Figure 184—Making hay from windrow to bale with a windrow pick-up press.

and is now generally considered necessary to the proper curing of legume hay. The hay loader is recognized as a laborsaver and cost-reducer that speeds up hay making. The hay press is gaining in popularity from year to year. While baling does not improve the quality of the hay, it does preserve its quality, makes it readily available for feeding or for sale. When we consider that loose hay in the stack or mow weighs from 4 to 5 pounds per cubic foot, and that hay baled under extreme pressure will weigh as much as 40 pounds per cubic foot, it is readily understood why baled hay is easier to store, handle, and transport, and why it feeds out with less waste. More recently, the windrow pick-up press (Fig. 184) has made it possible for the hay grower to cut, cure, and bale his crop without touching it with a fork. While the pick-up press is comparatively new, the side-delivery rake and the hay loader are widely used, and a more general knowledge of their value and their operation, care, and repair would be beneficial to farmers and farm workers.

Side-Delivery Rakes

When the hay is cut, the flow of ground moisture is shut off, but the plant is full of water. The problem, then, is to

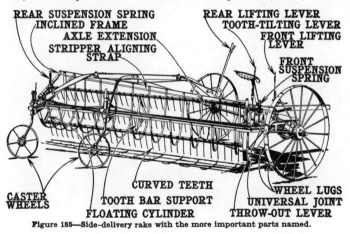

Figure 185—Side-delivery rake with the more important parts named.

reduce the moisture to a safe percentage for storing, and to do this in the shortest possible time.

The leaves, or tops, are left exposed to the sunlight, as they fall back over the mower cutter bar. If allowed to remain in this position very long, the leaves dry up and shatter. When this happens, the natural flow of moisture from stems to leaves is stopped and the moisture is "bottled up" in the stems. This results in unevenly cured hay.

The function of the side-delivery rake (Figs. 185 and 187) is to lift the hay from the swaths and place it in loose, fluffy windrows with the green leaves inside, protected from the sun's rays. The leaves, shaded by the stems, are cured rapidly by the free circulation of air through the windrows. They retain their fresh, green color and the stems are thoroughly cured for storing.

It is usually advisable to turn the windrow over at least once during the curing process to allow the bottom hay to cure properly. If

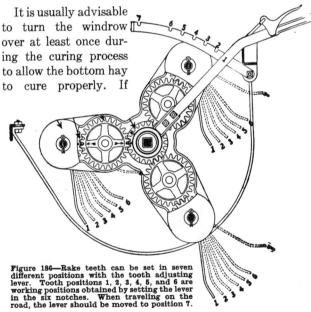

Figure 186—Rake teeth can be set in seven different positions with the tooth adjusting lever. Tooth positions 1, 2, 3, 4, 5, and 6 are working positions obtained by setting the lever in the six notches. When traveling on the road, the lever should be moved to position 7.

rainy weather catches the hay in the windrow, it is often necessary to turn it several times before it is thoroughly cured and ready for storing. The left front wheel is set in on the axle to allow enough of the reel to extend beyond, for turning the windrow upside down when the left-hand wheel is run next to the right-hand edge of the windrow.

Field Operation. Side-delivery rakes, once they are adjusted to suit field conditions, are easy to operate. The operator simply drives his team or tractor and oils his machine when necessary.

The most important adjustment is setting the teeth in the proper position, or angle, in relation to the surface of the ground. This is done with the tooth-adjusting lever, with which it is possible to set the teeth in six different working positions (see Fig. 186). The teeth should always be set as high as possible and still pick up all of the hay. This setting causes the curved teeth to pitch the hay, leaving the windrow as loose as possible, permitting free circulation of air.

In traveling on the road, the tooth-adjusting lever should be moved to notch seven. In this position, the teeth are

Figure 187—A field scene, showing two side-delivery rakes, hitched tandem, at work with a general-purpose tractor.

raised above the strippers out of danger of being bent by hitting obstructions.

The front lifting lever should be adjusted so that the front end of the reel is low enough to pick up the hay, but never so low that the teeth strike the ground. A trial in the center notch will usually give an indication as to the position in which it should be set.

The rear lifting lever is properly set when the rear end of the reel is slightly higher than the front end. This aids in making the windrow loose and fluffy.

Care Important. When starting a new side-delivery rake, or when using one that has been stored, it is a good plan to turn the reel by hand to be sure it revolves freely and that the teeth do not strike the stripper bars. Then throw the rake into gear and turn the wheel by hand to see that the tooth bars and gears work freely. Breakage of parts which results in serious delay can be avoided by taking these precautions before entering the field.

All wearing parts should be oiled regularly. An occasional thorough inspection for loose nuts, worn bolts, and other parts will add to the efficiency of the side-delivery rake.

Sulky Rakes

The sulky, or dump rake, shown in Fig. 188, is in common use in practically every section of the country. While it is easy to operate and adjust, many farmers work at a disadvantage when a slight adjustment would produce much better results.

The first requirement for good work is proper hitching. The rake shown in Fig. 188 is designed to work with the tongue 31 inches from the ground, measuring underneath at the front end. If this position is not maintained, the rake teeth will set at an improper angle, resulting in inferior work. If the tongue is too high, the teeth will have difficulty in clearing the hay after dumping; if too low, the teeth may fail to gather all of the hay.

Adjustments. Slight pressure on a foot trip lever causes the dump rods to engage in the wheel ratchets resulting in dumping of the rake. After the rake teeth have cleared the hay and started downward, they may be forced down quicker and held in position on the ground by pressure on the foot lever. An adjustment is provided at the hinge in this lever by which the wear can be taken up. If an adjustment is not made when the hinge becomes worn, the rake will be dumped with difficulty.

The height to which the teeth rise when the rake is dumped is controlled by adjusting a snubbing block bolt, located on the frame to the rear of the seat spring. If the rake rises too high and consequently does not get back to work as soon as it should, the block bolt must be screwed out of the block one or more turns. Turning the block bolt down permits the rake to rise higher when dumped.

If the rake repeats when it is dumped, the tension on trip spring is insufficient to hold the dump rod out of the wheel ratchets. More tension is produced by turning down the nut on the trip spring bolt.

When the wheel ratchets or dump rods become worn, the wheels and rods can be reversed, giving double wear.

Keep Nuts Tight. Because of the vibration attendant to raking, it is necessary that all nuts be kept tight. It is a

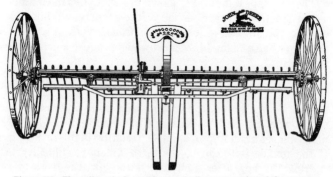

Figure 188—The sulky rake is used in practically every section of the country.

good plan to go over the rake at regular intervals for this purpose.

Oil, used liberally on axles and wearing parts, will make for good work and lengthen the life of a dump rake.

Hay Loaders

One of the greatest labor- and time-saving machines for the farmer is the hay loader.   It displaces hand-pitching—one of the hardest and most tiresome jobs on the farm—speeds up hay making, and cuts production costs.   The hay loader is needed on every farm where the hay is loaded in the field and put directly into the barns or hauled to stacks. Ten to fifteen acres of hay justify the purchase of a hay loader.

The ideal combination of equipment is the side-delivery rake and the hay loader. With these two, highest

Figure 189—Single-cylinder hay loader with carrier extension in lowered position.

quality hay can be produced at the lowest cost.

Figure 190—Building a load with the carrier of the loader down.

Types of Loaders. There are four distinct types of hay loaders—the single cylinder, the double cylinder, the combination raker-bar cylinder, and the raker bar. Each has its following and each has features that make it particularly adaptable to certain conditions.

The single-cylinder loader (Fig. 189) is used only for loading from the windrow. In handling hay from windrows, such as made by the side-delivery rake, it rakes its full width and delivers all of the hay onto the rack.

The floating gathering cylinder on the double-cylinder loader (Fig. 193) makes this type adaptable to loading hay from the swath as well as the windrow. When properly set,

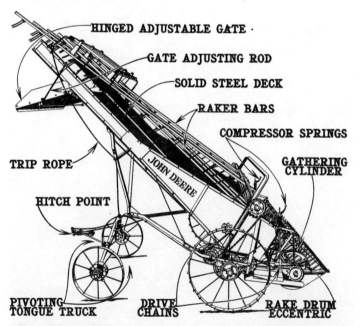

Figure 191—Combination raker-bar cylinder loader with the more important parts named.

it will do better work in rough fields than either the single cylinder or raker-bar loader. Its good work in picking hay from the swath makes it especially popular in some sections.

The combination raker-bar cylinder loader (Fig. 191) combines the most desirable features of both the double-cylinder and the raker-bar types, and threatens to displace the latter. The operation is similar to that of the double-cylinder loader. The floating cylinder rakes over its entire width, regardless of ground conditions, and the raker bar insures smooth elevation.

Raker-bar loaders operate on an entirely different principle than cylinder loaders. A series of bars to which are fastened malleable rakes elevate the hay from swath or windrow with a lifting or pitching motion, the result of an oblong stroke of the raker bars.

Operating Double-Cylinder Loaders. The most important factor in the satisfactory operation of a double-cylinder loader is the proper setting of the gathering cylinder. It does the best work when it is set in the highest position in which it will do a clean job of raking. If set too high, it

Figure 192—Completing a load with the carrier extension in raised position.

misses some of the hay; if too low, it gathers trash and the spring teeth scratch the ground, throwing dust into the hay. Height of the gathering cylinder is adjusted by cranks, one on each side of the loader. A spring placed behind each crank gives a floating action to the cylinder. In moving from field to field, the gathering cylinder should be raised by hooking the stop rods into the rear hole of the floating frame.

The loader should be hitched as closely to the rack as possible, but not so close as to cause it to strike the corners of the rack in turning.

The carrier should be kept fairly tight; that is, it should run with very little slack in the chains. Each season, before starting the loader in the field, it is a good plan to turn the carrier one complete revolution by hand to be sure slats register correctly with the driving notches on the drum.

The carrier extension can be lowered for starting the load (see Fig. 190) by releasing two levers. It is raised as the load goes higher by pushing up on the carrier bar. Fig. 192 shows the carrier extension in position for finishing the load. The operator must keep the extension in the same notch on both sides or the carrier will run crooked and cause trouble.

Single-Cylinder Loaders. With the exception of the discussion of the gathering cylinder adjustment, the foregoing information on the operation and care of double-cylinder loaders applies to single-cylinder loaders.

Care of Loaders. Because hay loaders have wood, light chain, and rope in their make-up, they should be stored in a dry place if possible. If the loader is permitted to remain in the open, the rain and sun shorten its life and increase upkeep costs. Shelter should be provided whenever possible.

Figure 193—Double-cylinder loaders are favored in many sections.

The usual admonition regarding thorough oiling of farm machines and keeping all nuts tight can be repeated for all hay loaders. Slack seasons furnish an opportunity to overhaul the hay loader along with other farm machines.

Hay Presses

The hay press shown in Fig. 194 is of the continuous type, a type in most common use throughout the United States today. In operation, the charge of hay, placed into the hopper by the feeder on the platform, is forced into the chamber by the feeder head. The plunger compresses the charge placed by the feeder and returns momentarily to its position while another charge is placed. Both the feeder head and the compressing plunger are operated by a large eccentric gear which provides maximum power on the compression stroke of the plunger and a quick return for the feeder head.

Grooved blocks, placed at regular intervals by the operator, determine the size of the bales. Baling ties are slipped through the grooves and fastened securely around the bale. As the bales are forced from the press, the blocks drop to the ground.

Operation and Adjustment. Before starting the press, inspect the entire outfit carefully for loose bolts. Oil all

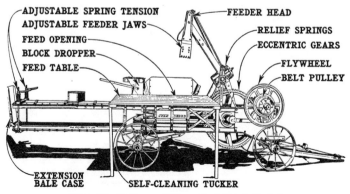

ADJUSTABLE SPRING TENSION
ADJUSTABLE FEEDER JAWS
FEED OPENING
BLOCK DROPPER
FEED TABLE

FEEDER HEAD
RELIEF SPRINGS
ECCENTRIC GEARS
FLYWHEEL
BELT PULLEY

EXTENSION BALE CASE
SELF-CLEANING TUCKER

Figure 194—Hay press with important parts indicated.

moving parts including main bearings for large gears, main shaft and intermediate gear shaft bearings, and crank-pin bearing. Turn the press by hand to be sure all parts are running freely. Adjust tension levers so that height inside the case is from 1/2 to 3/4 of an inch less than the height of a standard bale.

Extra care in feeding to be sure that small charges of uniform size (6 to 8 pounds of hay) are placed into the feeder will result in neat bales of uniform shape and size. When dropping division blocks, be sure that the block driver on nose of feeder head strikes the block squarely to insure positive drive of the block.

Baling ties should be reasonably tight, otherwise much of the work done in compressing the bale will be wasted.

In operation, all bearings should be oiled at regular intervals six to eight times daily. A daily inspection for loose bolts and worn parts is an important responsibility of the operator.

Questions

1. What qualities do you consider high-grade hay must have?
2. What legumes are grown in your community?
3. What is meant by "air-curing" hay? Is this method used extensively in your community?
4. What is the function of the side-delivery rake?
5. How should the teeth of a side-delivery rake be set in relation to the ground surface?
6. What are the principal adjustments necessary on a dump rake?
7. What is the advantage of reversible wheels and dump rods?
8. How large an acreage of hay justifies the purchase of a hay loader?
9. What four types of loaders are in common use? Which is most popular among farmers in your community?
10. How should the gathering cylinder be set for efficient work?
11. When is the extension carrier used and what are its advantages?
12. Why should hay loaders be put under shelter when not in use?
13. What are the advantages of baling hay for feeding or for market?
14. Why are eccentric gears used to drive the feeder head and plunger?

Part Five
POWER ON THE FARM

The adaptation of mechanical power and power-driven machines to farm work has been a developement of comparatively recent years. The tractor has made its greatest appeal to the farmer in the last fifteen or twenty years. And, within the same period of time, stationary and portable engines for belt work have come into common use on the farm.

The farm tractor has proved beyond a doubt its practicability for average farm conditions. It has aided materially in cutting production costs, increasing the working capacity of the farm worker, and speeding up farm operations. The remarkable increase in the number of tractors on farms within the past few years is but an indication of what may be expected in the future as tractors and tractor-operated machines are developed to an even higher state of efficiency and practicability. The prediction of engineering authorities that the tractor will eventually become the main source of power on the farm is already a reality on many farms.

One of the principal reasons for this trend to power farming is the fast-growing popularity of the general-purpose type of tractor. This latest developement in tractor design provides the farmer with power and equipment for doing all farm jobs—plowing, disking, planting, cultivating, mowing, and all other major operations. The need for horses is practically eliminated where tractors of this type are used. On many farms, the tractor is now the main source of power.

Farm engines have made farm life easier and more profitable. Many hard and tiresome jobs such as pumping water, running the cream separator or the washing machine, and cleaning grain, formerly done by hand, are now given to the stationary or portable engine. Grinding feed, running the concrete mixer, shelling corn, and other jobs for which power was often hired, are now handled at the farmer's convenience with his own engine or tractor furnishing the power.

Recent years have also brought a better and more general knowledge among farmers of the factors that govern the operation and care of internal-combustion engines. The result has been more satisfactory performance of farm power units and less expense for repairs and service.

Before attempting to operate a tractor or engine, the operator should make a careful and thorough study of the instruction books furnished by the manufacturer. Although the general principles that govern their operation are the same, each make of tractor or engine has somewhat different operating problems. The basic principles, common to practically all internal-combustion engines used on the farm, will be discussed in the following two chapters.

Internal-Combustion Engines. An internal-combustion engine is an engine in which the heat or pressure energy necessary to produce motion is developed in the engine cylinder, as by the explosion of a gas, and not in a separate chamber as in a steam engine boiler. The fuel, mixed with air, ignites, burns rapidly, expands inside the cylinder, pushes the piston back, turns the crankshaft, and so develops power. The power generated can be applied to the operation of machines through the belt pulley in the case of the engine, and through the belt pulley, drawbar, or power take-off in the standard type tractor.

The modern general-purpose tractor is equipped with a hydraulic power lift which furnishes a fourth power outlet used for raising and lowering integral equipment.

There are two general types of internal-combustion engines —two-stroke cycle and four-stroke cycle. The two-stroke cycle engine has a power impulse or working stroke every revolution. The four-stroke cycle engine burns its fuel charge every second revolution. There are four strokes of the piston from one power impulse to the next. All farm engines and tractors are of the four-cycle type. The four strokes of each piston in a four-stroke cycle engine are:

First: Suction or Intake. Here the piston draws a charge of fuel and air into the cylinder through the inlet valve.

Second: Compression. The piston, on its return, compresses the fuel and air mixture into the end of the cylinder, called the combustion chamber. Full power is secured only with good compression.

Third: Expansion or Power Stroke. At a point slightly in advance of full compression, an electric spark, produced by a magneto or battery, ignites the fuel. This causes a sudden high expansion pressure to act on the piston, pushing it back so that work is performed.

Fourth: Exhaust. On its return from the power stroke, the piston pushes the burned gases out of the cylinder, through the open exhaust valve and then through the exhaust manifold.

These events—suction or intake, compression, expansion or power, and exhaust—make the complete cycle.

Chapter XXI.
TRACTORS

The tractor is composed of five main units—the fuel, ignition, oiling, cooling, and transmission systems. Each is separate from the others, yet all are necessarily dependent upon one another. The efficiency of a tractor depends upon the proper functioning of all units—a condition which exists only when all units are properly cared for and properly adjusted.

For the discussion of this chapter, the tractor shown in Figs. 195 and 205 is used as a basis. It is a typical general purpose farm tractor having a two-cylinder, horizontal engine.

When a new tractor is delivered, it is ready to give efficient service for a long time, under normal conditions, without a great amount of adjustment. The operator's chief responsibility is in correct lubrication and proper care. However, when trouble arises, the operator should be capable of analyzing his machine and correcting the cause of his difficulty.

Fuel System. The fuel system is a vital unit of the tractor. Without fuel in proper quantity and in proper mixture with air, the engine cannot operate effectively. Distillate, kerosene, and gasoline are the principal fuels used in tractors. Distillate, sometimes referred to as furnace oil, fuel oil, tractor fuel, etc., is the least expensive of the three and is gaining in use among farmers whose tractors will burn it successfully. The tractors shown in this text operate on kerosene or distillate, using gasoline for starting.

A conventional type of carburetor, with an oil-wash air cleaner and a fuel filter, composes the main part of the fuel system. The air cleaner removes dust and grit from incoming air—all particles that would injure the cylinders and working parts are removed. The air is drawn into the cleaner through a high stack from such a level as to avoid the heavier, more harmful particles of dirt. The dust that passes through the stack is caught and retained in a mist of oil created by the draft of air drawn through the cleaner. Before each day's work, oil sediment cup at base of filter should be detached, the dirt-filled oil removed, and the entire cup washed in gasoline or kerosene to remove all the sediment. The cup should then be refilled to bead mark with new engine oil, and replaced.

A three-way valve between fuel tank and carburetor permits changing from gasoline, which is used for starting, to kerosene or distillate. When the engine becomes hot, the valve is turned to distillate or other low-cost fuel.

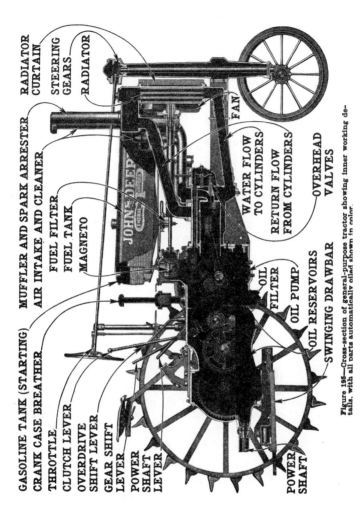

GASOLINE TANK (STARTING)
CRANK CASE BREATHER
THROTTLE
CLUTCH LEVER
OVERDRIVE
SHIFT LEVER
GEAR SHIFT
LEVER
POWER
SHAFT
LEVER

MUFFLER AND SPARK ARRESTER
AIR INTAKE AND CLEANER
FUEL FILTER
FUEL TANK
MAGNETO

RADIATOR
CURTAIN
STEERING
GEARS
RADIATOR

FAN

WATER FLOW
TO CYLINDERS
RETURN FLOW
FROM CYLINDERS
OVERHEAD
VALVES

OIL
FILTER
OIL PUMP
OIL RESERVOIRS
SWINGING DRAWBAR

POWER
SHAFT

Figure 195—Cross-section of general-purpose tractor showing inner working details, with all parts automatically oiled shown in color.

The amount of fuel used is controlled by adjustment of needle valves on the carburetor. Too much fuel is indicated by a black, smoky exhaust; too little, by a popping back through the carburetor.

If fuel does not flow readily, the filter bowl and the fuel strainer screens should be removed and cleaned. If the carburetor overflows or drips constantly, it is advisable to examine the float and the float valve seat—the float may be soggy or the valve seat dirty or worn.

The most important point in caring for and adjusting the fuel system is keeping out dust and dirt. Fuel should always be stored in clean containers, protected from dust and dirt; it should always be strained when filling the tanks. When engine trouble occurs, and it is certain a spark is being produced, look for the cause of the trouble in the fuel system.

Ignition System. The ignition system, producing a spark within the cylinder, causes the explosion of the fuel gases. Without a strong spark, timed to occur at the proper instant, the engine is difficult to start and will not operate efficiently.

The tractor shown in Fig. 205 is equipped with a high-tension magneto with impulse starter—a type in use on most tractors. It is properly aligned and timed when the tractor leaves the factory, but this setting may be disturbed by removing the magneto or tampering with the timing. In case timing has been disturbed and the engine's efficiency lowered, the magneto can be retimed, using the instruction book furnished with the tractor as a guide.

The impulse starter is an aid to starting. It consists of a coupling on the magneto drive shaft so arranged that, when set, it will revolve the magneto rotor quickly, so that no matter how slowly the flywheel is turned, a strong spark is produced. After the engine gains speed, the impulse starter automatically releases and the magneto rotor revolves normally.

One of the common causes of ignition trouble is dirty or improperly set spark plugs. Plugs should be cleaned regularly and adjusted if need be. There should be a space of .030 inch between the points of the plug for best results.

The magneto breaker points must be kept clean and adjusted to from .012 to .015 inch. If this setting is disturbed, difficult starting and irregular running will result.

The distributor contacts and brushes should be cleaned at regular intervals. A poor contact in the distributor will cause starting trouble, loss of power, and other difficulties. It is essential that all ignition parts be kept clean and free from dirt and excess oil. Magnetos that require lubrication should be oiled with a fine, light oil.

The spark lever controls timing of the spark in the cylinder. It is retarded when the engine is idling and advanced when the engine is pulling a load. Retarding the spark causes the explosion of the fuel to occur as the piston completes the compression stroke; advancing the spark causes an earlier explosion. The engine is capable of developing greater power with the spark advanced when pulling a load.

Figure 196—A general-purpose tractor with two-bottom plow.

Oiling System. Of all farm machines, the tractor requires the most careful oiling. Due to the nature of its work and the large amount of friction surface in its bearings and cylinders, the tractor must be properly lubricated with good oil and grease if it is to develop its maximum efficiency and last a normal length of time. No other factor affects the life of a tractor so greatly as does oiling.

The tractor of which a cross section is shown in Fig. 195 is provided with a positive-driven, gear-type oil pump which forces oil into the main bearings and through the drilled crankshaft to the connecting rod bearings, then through

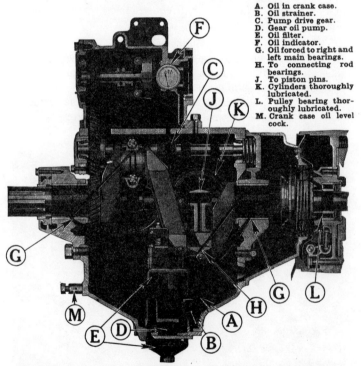

A. Oil in crank case.
B. Oil strainer.
C. Pump drive gear.
D. Gear oil pump.
E. Oil filter.
F. Oil indicator.
G. Oil forced to right and left main bearings.
H. To connecting rod bearings.
J. To piston pins.
K. Cylinders thoroughly lubricated.
L. Pulley bearing thoroughly lubricated.
M. Crank case oil level cock.

Figure 197—Cross-section of the oiling system showing how all working parts are automatically oiled. Color indicates oil.

holes in connecting rods to piston pins. A cross section of this system is shown in Fig. 197. The pistons and timing gears are lubricated by the oil spray thrown from the revolving parts.

When the engine is started, the oil indicator (see "F" in Fig. 197) will rise if the oiling system is working properly. If it does not rise, the operator should check the supply of oil in the crankcase. The trouble may also be in the oil-strainer screen or in the pressure relief valve. To insure lubrication, the indicator must be up when engine is running.

Every ten operating hours, the crankcase should be filled with fresh oil to the upper cock. Every sixty hours, it should be completely drained, washed with kerosene, and refilled with fresh, high-grade oil.

The gears, shafts, differential, and bearings in the enclosed transmission case operate in a bath of heavy oil. For proper lubrication of these parts, the transmission oil should not be permitted to go below the proper level recommended by the manufacturer of your tractor. In cold weather, when the oil becomes stiff, it should be thinned out with kerosene to a satisfactory consistency.

In addition to the transmission and crank case units, there are several oil fittings for high pressure gun grease that require regular attention. A thoroughly lubricated tractor will last longer and give better service than one that is given ordinary care.

Lubrication and adjustment of front wheel bearings should be given special attention. While the front wheels on most tractors are lubricated by means of a high-pressure fitting ("A", Fig. 198), it is advisable to remove the wheels occasionally to service the wheels.

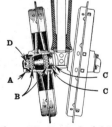

Figure 198—Front wheel detail.

To service front wheel, remove the wheel, take out all old grease, examine the bearings ("B", Fig. 198), clean the two felt washers

("C", Fig. 198) in gasoline, resoaking in transmission oil before replacing. If these washers are worn thin, replace with new ones. Pack hub and bearing with grease, replace wheel on spindle, and adjust as follows: relieve bearings of all weight by raising front end of tractor until wheels are free from the ground. Turn adjusting nut ("D", Fig. 198) tight; then, back off the adjusting nut 1/3 to 1/2 turn. Wheels should rotate freely but without end play. Lock adjustment at proper point by means of cotter key.

Cooling System. Tractor engines and other engines require a cooling system to hold the temperature at an even point best suited to insure complete combustion of the fuel and for efficient operation. Most engines adapted to farm use are water-cooled, although certain types of engines are successfully cooled by air.

The cooling system used on the tractor shown in Fig. 195 consists of a tubular radiator with a gear-driven fan. The circulation of the cooling water is thermosiphon- or temperature-controlled. When the cylinders warm up after starting the engine, the warmed water rises and is displaced by cooler water; the constantly rising warm water from the cylinders causes a circulation through the radiator where the water is cooled by a blast of air drawn through the radiator by the fan. In this manner, the motor is kept at an even temperature.

The radiator consists mainly of a core of vertical copper tubes attached to which are sheet-copper fins that form extra cooling area. As the fan draws a steady current of air through the radiator, the water is cooled as it flows downward. Fig. 195 shows a cross section of the cooling system.

A screen on top of the radiator tubes is designed to prevent foreign matter from entering and clogging the tubes. It should be kept clean at all times. Always use clean water and keep the level above the radiator tubes. Water should never be poured into an empty cooling system when the engine is hot.

Transmission System. The transmission system delivers the power from its source—the engine—to the drive wheels or the belt pulley where it is used to pull loads to operate belt-driven machines. It consists primarily of a clutch which connects and disconnects the power from the belt pulley and from the gearing through which the power is transmitted to the drive wheels. The clutch engages smoothly and allows the engine to pick up the maximum load gradually rather than with a jerk. A dry-disk type of clutch is used in the tractors illustrated in this text.

Power is transmitted to the belt pulley direct; the pulley is mounted on the engine crankshaft and is connected by the clutch which is a part of the pulley.

Power is transmitted from the engine to the drawbar through the clutch and through gears to the spline shaft. Two sliding pinions—one larger in diameter than the other—provide two tractor speeds forward. Either of these sliding pinions is moved into mesh with its respective gear that forms a part of the differential. On the general purpose

Figure 199—Standard type tractor with two-row corn picker.

tractors used in this text, two additional forward speeds are available by shifting an overdrive lever.   (See Fig. 205.)

Clutch Adjust-   Pulley Brake
    ing Nuts   Adjustment

Figure 200—The clutch is properly adjusted when all nuts
are drawn up to even tension.

A differential is a compensating gear unit that acts so as to transmit power to each of the drive wheels as the tractor turns a corner.   It permits the outer wheel to travel faster than the inner.

The clutch is properly adjusted when the nuts are drawn up to exactly the same tension, and the clutch operates with a snap, requiring some pressure to lock it.   If it is necessary to tighten the clutch, each nut must be turned down to the same tension, disregarding the number of exposed threads. (See Fig. 200.)

To replace clutch facings, remove dust cover and clutch adjusting disk.

When installing new clutch facings, put a dab of grease on the inside of first clutch facing to hold facing in proper position while clutch drive disk is being replaced.   Install second clutch facing in clutch adjusting disk, making sure to have the three short springs in place.   Adjust clutch as described above.

Differential Brakes.   The differential brakes, one for each drive wheel, are provided to facilitate turning right or left when working in row crops and to hold the tractor stationary on belt work.   They are easily adjusted to take up wear by tightening the adjusting screws.

Valve Adjustment.   The valves of an engine control the intake of fuel gases and the discharge of the burned gases from the cylinders.   Each cylinder has two valves—an intake and an exhaust valve.   The fuel is drawn into the cylinder through the intake valve, and the exhaust is discharged through the exhaust valve.

The valves are opened at the proper time by the action of tappet levers which are operated by cams driven from the cam shaft.   Springs close the valves when the tappets release their pressure on the valve stems.

Valve Grinding.   When we consider that the valves, when open, serve as gates to admit fuel or expel exhaust gases, and, when closed, serve to seal the combustion chamber against compression leaks, the need for a perfect seal between valve face and valve seat is apparent.   (See Fig. 201.)

Frequently, loss of power due to poor compression is caused by leaky valves.   Valves can be checked for leaks by turning the engine over against compression; if compression appears to leak through the valves, the valves should be ground and reseated.

Although the individual operations in grinding valves will vary with various types of tractors, the basic principles are the same.   For example, we will use the typical valve-in-head type of engine built into the tractor shown in Figs. 195 and 205.

Drain cooling system and remove cylinder head, using care to preserve the gasket between cylinder head and cylinder block.   If individual valves are not already marked, indicate their position by filing notches at the outer edge, or keep them in order of position as removed from the cylinder head or cylinder block. Remove valves by compressing valve springs and withdrawing valve spring cap and lock washer.

It is especially important to remove all carbon from valve heads, valve stems, valve guides, and cylinder head.

Place a light coil spring under the valve head, coat face of valve with coarse grinding compound, and slip valve into its proper port. With a brace or valve grinding tool, rotate the valve back and forth several times—do not turn the valve in one continuous direction. Relieve pressure at frequent intervals to permit valve to rise and reseat itself for repeated grinding. The grinding operation should be continued only until seat and face have a continuous, smooth seat of such width as specified by the manufacturer of the particular tractor being serviced. (See Fig. 201.) When valve is properly seated, wipe the valve face and seat clean, and finish the grinding operation with fine grinding compound. Clean the valve, valve-port, and valve-guide thoroughly to remove all grinding compound before reassembling.

Where valve seats have become so pitted that a smooth seat cannot be obtained by regrinding, it is necessary to provide new valve seats by means of special valve reamers such as shown in Fig. 202. The new seat is then carefully checked with the eccentrimeter, a precision gauge which indicates whether the surface of the seat is perfectly round or if irregular spots exist. In cases where the valve face has become warped or pitted, the valve must be refaced in a refacing lathe, removing only so much of the valve as necessary to true up the face. When these operations are necessary, it is best to have the work done by an experienced service man whose shop is equipped for work of this type.

In cases where valve guides have become so worn that they will not hold the valve in correct relation to the valve seat, the valve guides should be replaced. This is easily accomplished by driving the old valve guides out of the head with a hammer and a block of wood. The new guides should be installed in the same manner. Where a press is available, it will greatly facilitate this work. When new valve guides are installed, it is desirable to smooth up the valve seat in the head with the valve reamers, as the reamer pilot causes the

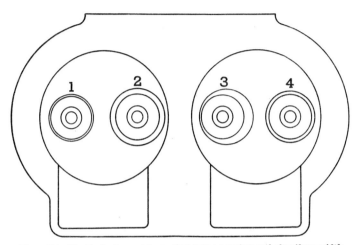

Figure 201—The ideal valve seat is one that forms a perfect seal of uniform width over the complete circumference of the valve-port. A seat too narrow (1) tends to cut or pound a groove in the valve, resulting in lost compression. With the seat too wide (2) it is almost impossible to get a perfect seal, and here again, loss of compression results.

Where the seat has been worn off center (3) due to worn valve guide or any other cause, it is impossible to obtain the proper seal between valve and seat which results in lost compression and uneven operation of the engine.

The properly seated valve (4) forms the perfect compression seal, and insures proper engine performance for the longest period. Width of seat varies with power and type of engine. See manufacturer's instruction book or consult your dealer for exact width for your engine.

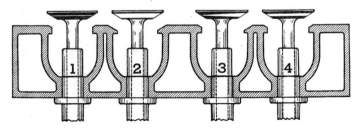

face to be made in cylinder head in correct relation to the valve guide.

When cylinder head is put back on tractor, apply shellac or grease to both sides of gasket and place smooth side next to cylinder. Draw nuts up uniformly and tighten thoroughly.

The valves must be timed to open and close at exactly the proper moment. Adjustment is provided in the tappet adjustment screws which should be set according to directions furnished by the manufacturer.

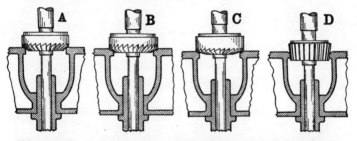

Figure 202—The "roughing" reamer, "A", is used first to remove the hardened, glazed surface from valve seat, after which the seat is smoothed by the finishing reamer, "B". The seat must now be reduced to proper width. Using reamer, "C", the seat is narrowed from the top; reamer, "D", is then used to narrow the seat at the base to the width specified by the manufacturer of the tractor.

Bearings. The connecting rod bearings can be adjusted to take up wear with little difficulty. Remove crank case cover, or pan, turn crankshaft until bearing is easily accessible, and remove bearing cap and block of laminated brass shims as shown in Fig. 204. With a pocketknife, remove one layer (.003-inch) from each shim which is usually sufficient to take up wear. (See Fig. 203.)

In reassembling the bearing, be sure laminated shims are placed between the two steel shims before bearing cap is replaced. Draw nuts up tight, put cotter pins in place, and spread ends to lock in position.

The bearing should fit snugly, but not too tight, when the cap is replaced and the nuts tightened. It may be necessary to repeat the operation to get the bearing to fit properly.

The main bearings are taken up in the same manner as the connecting rod bearings.

Power-Shaft Attachment. A power-shaft attachment is a third means of taking power from the tractor to drive such machinery as the tractor binder and corn picker. It drives the entire mechanism of this type of machine; the bull wheel acts merely as a supporting or carrying member.

Figure 203—Laminated brass shims provide for taking up wear in bearings.

The attachment consists, mainly, of a shaft, extending back from the tractor, that is driven by the regular transmission. It is especially advantageous in fields where traction conditions make it difficult for the bull or drive wheel of a machine to develop enough power to drive its mechanism.

Figure 204—Removing the connecting rod bearing cap.

Power-Lift Attachment. The power-lift attachment, which supplies power for raising and lowering the working equipment, is of the hydraulic type, the pump of which is driven from the power shaft. The principles of its design and construction will be considered more fully the following chapter.

Care in Handling. The tractor is more responsive to careful handling than any other machine. It lasts longest, does its best work, and delivers its power at the lowest cost when given the best care. The operator who gives more than usual attention to proper oiling, fueling, and cooling of his tractor will be more than repaid in better performance.

Questions

1. What changes have been brought about on the farm by the more general use of tractors and farm engines?

2. What is an internal-combustion engine?

3. What is the difference between a two-stroke cycle engine and a four-stroke cycle engine? What are the four strokes of the latter?

4. What are the five main units of the tractor, and what is the function of each?

5. Describe the fuel system. Of what use is the oil-wash air cleaner?

6. Why is distillate or kerosene preferred as tractor fuel? How is the quantity of fuel used controlled?

7. What is the purpose of a magneto? The impulse starter?

8. What is the proper setting of spark plug points?

9. Describe the oiling system of the tractor illustrated. How often should the crank case be drained?

10. Why is a cooling system necessary?

11. Describe how the power generated is transmitted to the drive wheels; to the belt pulley.

12. What is a differential unit? Describe a clutch.

13. What is the function of the valves?

14. How would you take up the bearings?

15. What are the advantages of a power-shaft attachment?

16. What per cent of farms in your community have tractors, and what per cent should have them?

General Purpose Tractors

Modern farm tractors may be divided into two general classes: the general purpose type; and the standard tread type. While the standard tread tractor is of earlier design, the general purpose type has come into such universal favor that it will be given first discussion in this text.

The general purpose type of tractor, as its name implies, furnishes power for practically all farm work. Not only does it perform all of the drawbar, belt, and power shaft jobs, but, in addition, puts speed and economy into the planting and cultivating of row crops.

Taking into account the size of farms, the nature and relative importance of the various jobs to be done, and all other considerations, practical experience has settled on a tractor of sufficient power to handle the six-horse load as being best adapted to general purpose use on the majority of farms. Smaller general purpose tractors, built to handle the

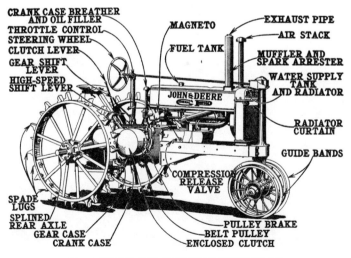

Figure 205—Adjustable tread general-purpose tractor.

four-horse load, are gaining in popularity on smaller farms as a supplementary power on farms where larger tractors are used. In addition to these types, there are variations of the general purpose tractor adapted to special work, such as the tractor shown in Fig. 206 with adjustable front axle for use in extremely soft ground, and the tractor with single front wheel (Fig. 215) used for work in garden crops where rows are extremely narrow.

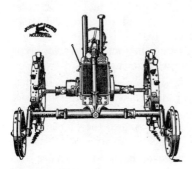
Figure 206—General purpose tractor with adjustable front axle.

In heavier soil and on larger farms where row-crops are raised, the general purpose tractor of three-plow power fills the need for a tractor adaptable to the varied operations in plowing, planting, cultivating, and harvesting, in addition to belt work.

Figure 207—Adjustable tread tractor planting four rows of cotton.

Figure 208—Planting corn with a four-row towed-type tractor planter.

Weight Is Factor. A general purpose tractor must be heavy enough to give good traction efficiency in plowing and similar heavy work, yet no heavier than needed, because a larger part of its work is on mellow soil.   Similarly, its engine must have enough power for the heavier drawbar jobs, yet be efficient at lighter loads.   The clearance of all parts that

Figure 209—Cultivating cotton with a two-row tractor cultivator.

Figure 210—Cultivating corn four rows at a time with an adjustable tread tractor.

pass above cultivated plants must be sufficient to allow tractor to pass over them without harm, yet the machine must not be top heavy.

Adjustable Tread.   In most sections of the country, the adjustable tread type tractor is gaining in popularity over

Figure 211—General purpose tractor with power-driven mower attached.

the standard tread type. This is especially true where crops are grown in narrow rows. Cotton, potatoes, and beans are representative crops usually planted in narrow rows and which are better adapted to cultivation with adjustable tread tractors than with the standard tread type. When set for cultivating, the adjustable tread straddles two rows of regular or narrow width, while the wheels of the standard tread run rather close to the rows when straddling one row in narrow-row crops.

A typical general purpose tractor, with adjustable tread, which can be equipped for a wide variety of uses in almost any row crop is illustrated in Fig. 205. Two- and four-row planters, and two- and four-row cultivators for cotton, corn, and other crops, two- and four-row bedders for cotton are some of the equipment that can be used with this tractor. For such jobs as plowing, the rear wheels can be reversed right to left and set in 56-inch tread which largely overcomes side draft. Fig. 212 shows a detail of rear hub and splined axle shaft which makes possible the variation in wheel tread shown in Fig. 213. Figs. 207 and 210 show

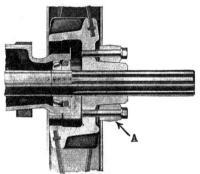

Figure 212—Wheel hub of general purpose tractor, showing drive wheel hub clamp, "A", which locks wheel to splined axle.

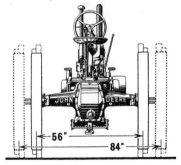

Figure 213—Diagram showing maximum variation in rear wheel tread.

the four-row planting and four-row cultivating units attached to the adjustable tread tractor illustrated in Fig. 205. Manufacturers of tractors and farm equipment now provide a wide variety of equipment for their tractors, making it possible to grow and harvest practically any crop, using tractor power exclusively. The attachments and machines available are so numerous as to make impractical a complete consideration of them in this text. The implement dealer's store provides the best means of seeing and studying the various equipment available for each community.

Clearance Important. The general purpose type of tractor must be constructed to allow all necessary clearance above the growing crops. Ample clearance is gained in the adjustable tread general purpose tractors by several im-

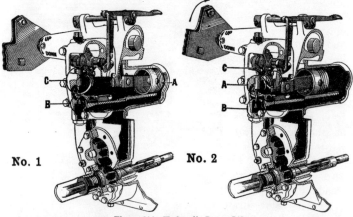

Figure 214—Hydraulic Power Lift.

In No. 1, the power lift is shown in the lowered position. The piston "A" is extended and the rocker shafts are rotated so that the implements are in their lowered position. The lower valve "B" and the cylinder type valve at the top "C" are open, permitting the oil to recirculate.

No. 2.—When the foot pedal is pressed down, the upper valve "A" closes, while the lower valve "B" opens. This directs the flow of oil against the head of the piston forcing it to travel back in the cylinder, causing rocker arms to rotate, lifting the equipment. As the piston approaches the end of the stroke, the connecting rod contacts the trip lever "C" which rotates the ratchet, keyed to the cam shaft.

portant features of construction. By mounting the front of the tractor on a single support and extending rear wheel tread to straddle two rows, the engine is placed between the rows. In addition, the high drive wheels, in combination with the properly designed rear axle housing, provide ample clearance for cultivating all row crops.

It is highly desirable in planting and cultivating to turn completely around without stopping and be in position to continue back on the next set of rows. To make this possible, there is a separate brake for each rear wheel on the general purpose tractor shown. Pressing the brake pedal for the inner wheel holds the wheel back and aids the front wheels in swinging the tractor sharply around.

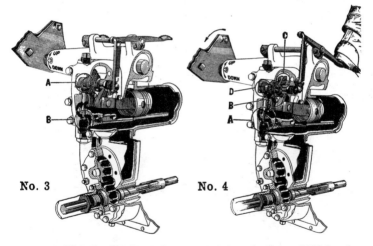

No. 3—When the piston has reached the end of the stroke, the cam "A" is forced off the valve by the ratchet and connecting rod. This closes the lower valve "B" and permits the oil to recirculate. The piston will now stay in the raised position until the foot pedal is pressed down.

No. 4—When the pedal is pressed down, both the upper and lower valves remain partially open "A". The oil can now return from the head of the piston to the reservoir. Note that the pawl "B" does not immediately engage in the ratchet as the foot pedal is pressed down. It is necessary that the pedal remain pressed down until the piston has traveled a sufficient distance to permit this engagement as shown in "A". The distance which a foot pedal can be pressed down is regulated by the stop "C". If the pedals contact the power lift housing, it will be difficult to rotate the cam "D" to the proper position.

Hydraulic Power Lift. To assist further in making the turns at the ends of the rows without delay, there is built into these tractors a hydraulic power lift device, which is connected to the planter, cultivator, or other equipment, and operated by a double control pedal which enables the operator to use either foot in raising and lowering the integral equipment. When the control pedal is touched, the working equipment is lifted promptly by engine power. A second touch of the pedal releases the lift and allows the equipment to drop to work against a cushion of oil in the lift. An adjustment is provided whereby the operator may vary the resistance of the oil cushion in the lift to control speed of drop for both light and heavy equipment. Fig. 214 illustrates the action of the hydraulic power lift in raising and lowering. Besides being independent of the operator's physical strength, the engine-driven power lift is independent of the forward movement of the outfit—lifting or lowering may be done while the tractor is stationary as well as when moving.

Figure 215—Cultivating six rows with a small general purpose tractor and integral cultivator.

In a general purpose tractor, flexibility of speed has much to do with capacity and efficiency. In cultivating, especially, there are times when it is desired to go very slowly. At other times, both speed and effectiveness are gained by traveling fast and throwing the soil briskly. To meet this wide range of speed demands, it is usual to provide several forward speeds in the transmission gears. These speeds in the general purpose tractor used for our example are 2-1/3, 3, 4-3/4, and 6-1/4 miles per hour. Still further variation in speed may be had by throttling down the engine.

The power take-off device, which supplies power directly by shaft to machines being pulled by the tractor, has found wide application and great usefulness on tractors of both the standard and the general purpose types.

Care Important. When the general purpose tractor supplants animal power on the farm, it is doubly important that it be properly cared for. If the owner is dependent upon his tractor for all farm jobs, delays are costly. Careful handling, strict attention to oiling, adjusting, and repairing the tractor and the equipment that is used with it will result in greater satisfaction and greater net profits.

The care and operation of the various units attachable to the tractor are discussed in the chapters devoted to each particular type of machine.

Standard Tread Tractors

While the general purpose tractors, described on preceding pages, meet the needs of the row-crop farmer in plowing, planting, cultivating, and harvesting his crops, the particular power requirements of the small-grain grower and the orchardist are best met by tractors of standard design, especially adapted to the work at hand. On larger farms where row crops are grown, standard type tractors are often used to supplement the general purpose tractors in preparing seed beds and harvesting the crops.

What has been said about the design and operation of the general purpose tractors on the foregoing pages applies so

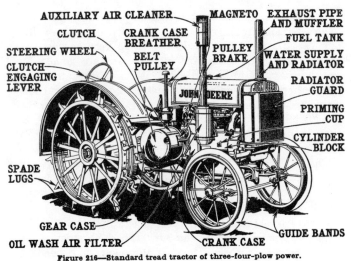

Figure 216—Standard tread tractor of three-four-plow power.

fully to the standard types that a further discussion is unnecessary.

The standard tread tractor, furnishing power at three outlets, the drawbar, belt, and power take-off, is used for practically all power requirements except planting and cultivating. The tractor shown in Fig. 216 is a typical three-four-plow tractor of this type. For smaller farms, a standard type tractor of two-plow power shown in Fig. 217, is available. A further variation of the standard type is the

Figure 217—A standard tread tractor of 2-plow power.

orchard tractor, a tractor of two-plow power, built low and compact, and with wheels and pulley shielded to facilitate operation under low-hanging limbs in orchards. Fig. 218 shows an orchard tractor at work in an English walnut grove.

Questions

1. *Name and describe two types of tractors.*

2. *Tell some of the advantages of the general purpose tractor.*

3. *Why is weight an important factor in the general purpose type of tractor?*

4. *What are the main differences in construction between the general purpose type and the standard tread type of tractor?*

5. *What are the advantages of a power lift on a tractor? Describe the principle of its operation.*

6. *Of what use are the wheel or differential brakes on the tractor?*

7. *Why is variation of speed a valuable asset in a tractor of this type?*

8. *What are the important advantages of the standard type tractor?*

9. *Review the more important points in the care of tractors.*

Figure 218—Orchard-type tractor working in an English walnut grove.

Chapter XXII.

FARM ENGINES

In most agricultural sections, the majority of farms have one or more engines that handle any number of power jobs. This situation requires a general knowledge among farmers of the principles that govern engine operation.

The average engine is not difficult to operate, and its adjustment is mainly a matter of using good judgment. The following text is designed to give the student the basic facts necessary to an understanding of engine operation and adjustment, using the engine shown in Fig. 219 as a basis. Other farm engines are similar in design, and operate on the same principles.

The engines shown in this chapter are of the four-stroke cycle, water-cooled, internal-combustion type. Practically all engines, with the exception of marine engines, are of this type.

The main points in the operation and care of stationary and portable farm engines are not radically different from those given for the engine unit of the tractor described in the preceding chapter.

The engine is composed of four units—the fuel, oiling, ignition, and cooling units. The perfect functioning of each is necessary to a smooth-running, powerful engine. Fig. 220 shows a cross-sectional view of a gasoline engine with the name and description of each of the important working parts beneath the illustration.

Fuel System. The fuel system is simple, yet it may be the cause of much trouble if a few simple rules are not followed. Only clean fuel should be used, the operator being careful that dirt or dust does not get into the supply tank when the tank is being filled.

Nos. 9, 11, 12, 13, and 14, in Fig. 220, show the important parts of the fuel system. When starting the engine, the air shutter (see Fig. 221) is closed to cause the intake stroke of the piston to fill the fuel line with gasoline and the combustion chamber with fuel vapor.   The check valve, No. 11, Fig. 220, holds the fuel at the mixer needle while the engine is operating.   The mixer needle controls the amount of fuel entering the cylinder.

If the engine does not get gasoline, the fuel line and the strainer should be removed and cleaned.   Trouble may also be caused by the check, if a piece of dirt lodges beneath it and permits the fuel to flow back into the tank.   The check must hold the fuel line full of gasoline.

Ignition.   Gasoline engines are operated either with a magneto or battery.   The former is preferred, as the maintenance cost is lower.

The magneto is properly set when it leaves the factory and should give a long period of service without much adjustment.   Most important of all is proper timing, which may be disturbed by removing the magneto or by moving the trip bracket unintentionally.

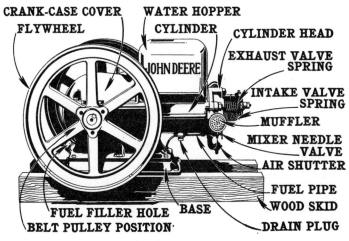

Figure 219—Typical style of gasoline engine with parts named.

The igniter must trip to produce the spark within the cylinder when the mark "spark" on the flywheel is level with or

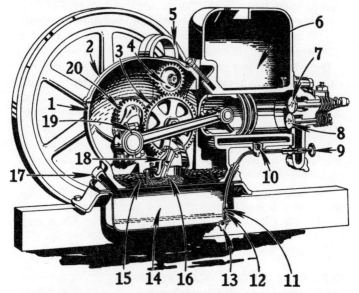

Figure 220—Cross section of a gasoline engine, parts and their purposes being described as follows:

1. Crank case contains all important operating parts.

2. Crank case cover prevents dirt or foreign matter from getting into operating parts. Removable without disturbing magneto or other operating parts.

3. Cam gear drives cam shaft which governs ignition and valve timing. It also drives magneto and governor.

4. Magneto gear drives magneto at the same speed and same direction as crankshaft.

5. Magneto furnishes hot, fat spark for starting and continuous running.

6. Water hopper.

7. Exhaust valve.

8. Intake valve.

9. Mixer needle valve for adjusting amount of fuel.

10. Drain plug for removing water.

11. Check in fuel line keeps gasoline at mixer.

12 Strainer screen prevents foreign matter from entering check, fuel line, or needle valve.

13. Drain plug for flushing water and sediment from fuel tank.

14. Gasoline tank.

15. Oil reservoir—capacity for many hours' continuous running. Note maximum oil level shown.

16. Oil pan—revolving governor (18) splashes oil to all parts of crankcase thoroughly lubricating operating parts. Oil pan governs oil supply.

17. Large gasoline filler hole. Convenient oil-filler on opposite side of engine. (See Fig. 221.)

18. Governor runs in oil.

19. Connecting rod bearings die-cast, removable and replaceable. Metal shims for adjustment.

20. Main bearing fitted to crank case. Replaceable die-cast bearings with metal shims for adjustment.

slightly above the exhaust rod (see Fig. 221) and the exhaust rod is clear back toward the flywheel, just starting ahead. If the igniter trips before this point, loosen the clamp bolts that hold the trip bracket and adjust it back toward the flywheel. If it trips later than this, adjust the bracket ahead. The face of the igniter trip must set flat against the igniter hammer when the bolts are again tightened.

When using batteries and a coil for ignition, timing is adjusted in the same manner. The coil should buzz when "spark" on the flywheel is even with the exhaust rod.

The coil points should be set 1/32 inch apart when the spring is held down.

A dirty or improperly adjusted spark plug may cause starting or operating troubles. To test, the plug should be removed and touched against the cylinder. When the coil vibrates, the plug should show a good spark. If the spark is weak, clean the plug and adjust the points to spacing recommended for your engine.

Figure 221—Showing the magneto side of gasoline engine with side plate removed.

When using batteries for ignition, the operator must be certain that the switch is left open when the engine is not running.

Speed Adjustment. The operating speed of the engine shown in Fig. 219 is controlled by the speed change nut (see Fig. 221).   It can be set to vary the speed from one-half to the full rated speed.

The oiling system on most farm engines is usually simple and requires little attention.   Fig. 220, with its accompanying descriptive matter, explains the oiling system used in the engine illustrated.

Size for Every Need. Farm engines are built in several sizes to conform to different requirements.   The engine

Figure 222—A stationary engine of the heavy-duty type.

illustrated in Fig. 219 is built in 1-1/2-, 3-, and 6-H.P. sizes, offering an economical power plant for every use.

The smaller sizes of engines can be used either as stationary engines or mounted on a hand truck. The latter permits their transportation from place to place about the farm, making them more useful. The heavy-duty engine, designed to burn distillate and similar low-cost fuels, is available for the heavy jobs. Fig. 222 illustrates a stationary engine of this type.

Farm engines develop greater efficiency and last longer when given more than ordinary care.

Questions

1. *What four main units does the gasoline engine have?*
2. *What are the important points to remember about the fuel system?*
3. *How would you adjust the igniter to trip at the proper time when using magneto? Batteries?*
4. *How is speed controlled on the engine illustrated?*
5. *What determines the size of engine a farmer should buy?*

224

Part Six
SOIL FERTILITY

Soil is the source of food that sustains mankind. It is productive so long as it contains sufficient quantities of all the essential plant-food elements, and so long as the right methods are observed by the farmer in working his land.

The supply of plant-food elements in the soil is not inexhaustible. Like a bank account, it becomes depleted if the amount withdrawn is greater than the amount deposited.

With the removal of each crop, the soil surrenders some of its fertility. If an equal amount of fertility is returned by man, productiveness is maintained.

Chapter XXIII.
MANURE SPREADERS

Experience has proved to farmers in every section of the country that barnyard manure is of great value as a soil fertilizer and a factor in permanent agriculture. The insistent urgings of scientists and farm experts have moved farmers to try a regular plan of covering their fields with the manure and waste vegetable matter from their barns and feed yards. The results have proved gratifying, and, as a general rule, farmers value highly the manure that was once considered a useless by-product of farming that was to be disposed of in the easiest possible manner or permitted to rot and waste away in piles about the barnyard.

One of the major reasons for the wastage of manure in the early days of agricultural expansion was the great amount of hand labor required to get it distributed evenly over the fields. The hard work of pitching into high wagon boxes, unloading into piles, and spreading by hand or spreading direct from the load was distasteful even to the farmer who was most conscientious about maintaining the fertility of his fields. The result was a more or less general laxity in conserving the manure that is now valued so highly.

The following graphs, furnished through courtesy of the Ohio State University, supply interesting facts about the value of manure in maintaining soil fertility, and, in addition, show results obtained from various methods of handling manure.

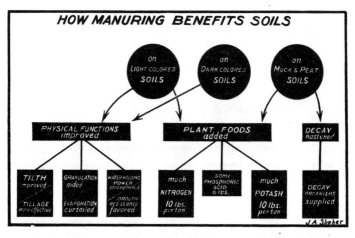

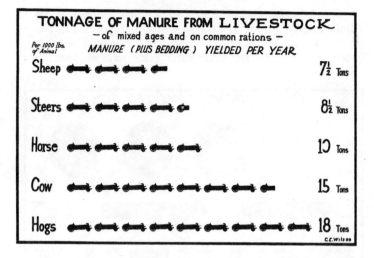

TONNAGE OF MANURE FROM LIVESTOCK
– of mixed ages and on common rations –
Per 1000 lbs. of Animal *MANURE (PLUS BEDDING) YIELDED PER YEAR*

Sheep   7½ Tons

Steers   8½ Tons

Horse   10 Tons

Cow   15 Tons

Hogs   18 Tons

C.E.Wilson

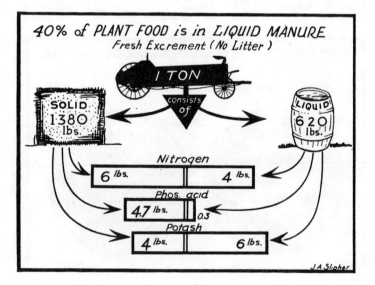

40% of PLANT FOOD is in LIQUID MANURE
Fresh Excrement (No Litter)

1 TON

consists of

SOLID 1380 lbs.   LIQUID 620 lbs.

Nitrogen
6 lbs. 4 lbs.

Phos. acid
4.7 lbs. 0.3

Potash
4 lbs. 6 lbs.

J.A.Slipher

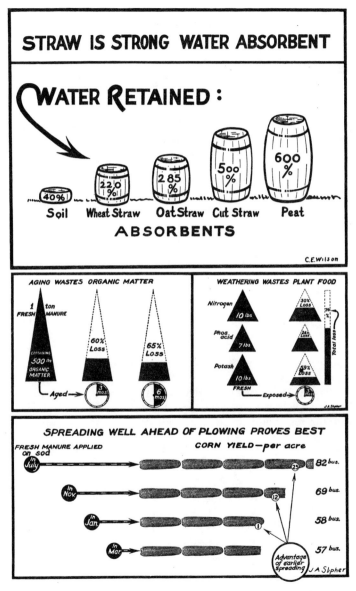

STRAW IS STRONG WATER ABSORBENT

WATER RETAINED:

| 40% | 220% | 285% | 500% | 600% |
| Soil | Wheat Straw | Oat Straw | Cut Straw | Peat |

ABSORBENTS

C.E.Wilson

AGING WASTES ORGANIC MATTER

1 ton FRESH MANURE

CONTAINING 500 lbs ORGANIC MATTER

60% Loss — Aged — 3 mos.

65% Loss — 6 mos.

WEATHERING WASTES PLANT FOOD

Nitrogen 10 lbs — 30% Loss

Phos. acid 7 lbs — 24% Loss

Potash 10 lbs — 49% Loss

FRESH — Exposed —

Total loss

J A Slipher

SPREADING WELL AHEAD OF PLOWING PROVES BEST

FRESH MANURE APPLIED   CORN YIELD—per acre

In July .... 25 .... 82 bus.

In Nov. .... 12 .... 69 bus.

In Jan. .... 1 .... 58 bus.

In Mar. .... 57 bus.

Advantage of earlier spreading

J A Slipher

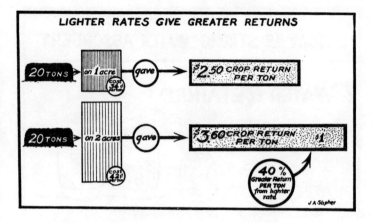

LIGHTER RATES GIVE GREATER RETURNS

Mechanical Spreaders. The introduction of the manure spreader not only gave the farmer an easier and quicker method of spreading his barnyard manure, but it also paved the way for a more concentrated effort on the part of agricultural leaders to impress upon him the value of its use in increasing crop production. Winning the farmer to the use of the spreader was comparatively easy when both the labor-saving and crop-producing features were pointed out to him.

Today, the great majority of farms are equipped with some type of manure spreader. Farmers are adding to their profits and building up their soils by utilizing the manure that was once wasted. The general practice is to spread the manure on the fields as it accumulates, thus getting the full benefit of all the plant-food elements.

To be most effective, manure must be spread evenly over the entire surface of the field. If it is deposited in bunches, part of the soil is without fertilizer and part is oversupplied. If the manure has a large amount of straw in it, difficulty is experienced in plowing and cultivating spots where it is bunched. The proper loading and operating of the manure spreader will overcome or lessen the possibilities of uneven spreading, provided the spreader is constructed properly.

Types of Spreaders. There are two general types of manure spreaders—horse-drawn, in which the traction of the rear wheels furnishes power for driving the beaters, and tractor-driven, in which the tractor engine, through the power take-off, operates the beaters.

Figure 223—Tractor spreader in which power for driving the beaters is furnished by the tractor engine through the power take-off.

A style of horse-drawn spreader in common use is shown in Fig. 224. It has three beaters. The upper and main beaters shred the manure; the spiral beater deposits it evenly over the entire width, making a well-defined line beyond the drive wheels.

The manure is carried back to the beaters by a steel slat conveyor, the speed of which is controlled by the feed lever

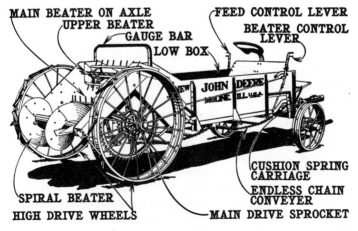

MAIN BEATER ON AXLE
UPPER BEATER
GAUGE BAR
LOW BOX
FEED CONTROL LEVER
BEATER CONTROL LEVER
CUSHION SPRING CARRIAGE
ENDLESS CHAIN CONVEYER
SPIRAL BEATER
HIGH DRIVE WHEELS
MAIN DRIVE SPROCKET

Figure 224—A horse-drawn manure spreader.

from the driver's seat. From five to twenty loads can be spread per acre, according to the setting of the feed lever.

The operator must be sure to keep the feed lever forward, in neutral, whenever the machine is not in gear or whenever the beaters are not operating. If the feed lever is left in operating position when starting to the field with a load, the conveyor forces the load back against the beater, resulting in breakage in some part of the feed mechanism. The feed lever should be thrown into neutral also when turning sharply while spreading.

With the control lever, the operator shifts the main drive chain so that it is in contact with the large drive sprocket. The three beaters are driven by two chains, both of which are set into action by the drive sprocket. The beaters should not be put into gear in this manner while the machine is in motion. The control lever should be moved to the rear only when the spreader is standing still.

In care and operation, the tractor-driven spreader (Fig. 223) is so similar to the horse-drawn type that the instructions given for the horse-drawn apply to the power-driven spreader as well.

Building the Load. It is much easier on both the team and the spreader if the operator starts to load from the front end, finishing at the beater end. The shredding process, which is the work of the beaters, is less of a strain when the load is built in this manner, resulting in lighter draft and less wear on the machine.

Questions

1. What relation has soil fertility to the production of food?
2. Name the types of fertilizers used on farms in your community.
3. What are the advantages of using a manure spreader over hand-spreading methods?
4. How should manure be spread to be of most value?
5. What are the two types of spreaders?
6. How would you build a load of manure for best results?
7. How is the quantity to be spread per acre controlled?
8. Why is thorough oiling important?
9. What per cent of farms in your neighborhood are equipped with manure spreaders?

Chapter XXIV.
LIME AND FERTILIZER SOWERS

In every section of the country, there are soils that would produce better crops with the application of lime, the proper commercial fertilizer, or a combination of both. Sour or acid soils that have been depleted of their lime by constant cropping or poor drainage can be rejuvenated and their productivity greatly increased by the application of lime. "Worn out" soils can be brought back to a productive state with commercial fertilizers and lime applied in correct amounts. The lime sower affords the easiest and most economical method of distributing these materials.

Lime and fertilizer distributors are made with two types of feeds—the star force feeds that handle from 50 to 5000 pounds per acre, and the rotary wing feeds that sow from 200 to 8000 pounds per acre. This latter type, when equipped with pneumatic tires, is used for distributing calcium chloride, cinders, salt, and other materials on roads.

Spreads Evenly. Uniform spreading of the correct amount of material per acre is the first requisite of a good lime sower. Bunching or skipping brings unsatisfactory results.

Figure 225—Spreading a uniform layer of lime with a modern lime and fertilizer distributor.

Fig. 225 shows a lime and fertilizer sower which spreads lime or fertilizer evenly in any amount from 200 to 8000 pounds per acre. The operator's only responsibilities are filling the hopper, setting his machine to sow the desired amount per acre, and driving his team. Two levers on the rear of the hopper (see Fig. 225) provide adjustment for quantity to be distributed.

Agitator Keeps Material Flowing. A revolving agitator in the bottom of the hopper keeps the lime or fertilizer flowing evenly through the feed openings. Its purpose is to prevent clogging or bridging of the material and the consequent skipping that would result.

Scattering boards, hung beneath the feed openings, aid in even distribution. The material is deflected and spread as it falls from the feeds. These boards also aid in more even distribution on windy days. Adjustment up or down is provided by a chain on each board.

If properly cared for, a lime sower should last for many years and prove to be a profitable investment.

In addition to lime and fertilizer distributors, a simple lime-spreading attachment for spreading lime in any quantity from 3/4 of a ton per acre and up is available for many modern manure spreaders, thereby making the manure spreader doubly useful in preserving soil fertility.

Questions

1. *What are the advantages of using a lime or fertilizer sower? Discuss methods of sowing.*

2. *Why is even distribution important?*

3. *What is the purpose of the agitator?*

4. *Why is a scattering board used?*

5. *Are lime and fertilizer sowers used in your community?*

Handy Acreage Chart

Copyright Deere & Co.

With the aid of this Handy Acreage Chart, the acre-per-hour capacity of machines of any width cut at various speeds can easily be determined.

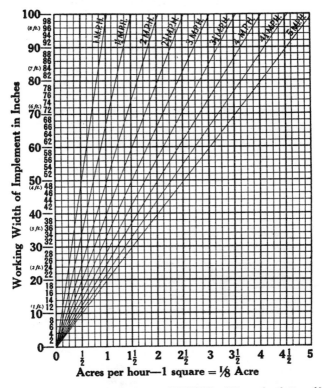

Acres per hour—1 square = ⅛ Acre

DIRECTIONS: In the left-hand column, find the line that represents the working width of your equipment. Follow this line to the right until it touches the diagonal line representing the speed of travel. Follow vertical line to the bottom of chart and estimate hourly acreage from nearest figure. In figuring acreage for implements wider than 100 inches, figure as above for half the width and multiply the result by two.

EXAMPLE: (Using a three-bottom, 16-inch tractor plow cutting 48 inches, traveling at 3-1/4 M. P. H.) Follow the line numbered 48 to a point midway between the diagonal lines marked 3 M. P. H. and 3-1/2 M. P. H., which represents the speed at which you are traveling. From this point, drop down to the bottom of the chart. Acreage covered is just a trifle less than 1-5/8 acres per hour.

MILES TRAVELED IN PLOWING AN ACRE

Width of Furrow, Inches		Miles
10	....	9-9/10
11	....	9
12	....	8-1/4
13	....	7-1/2
14	....	7
15	....	6-1/2
16	....	6-1/6

ACREAGE PER MILE OF VARIOUS WIDTHS

Width	Acres	Width	Acres
1 foot	0.121	15 feet	1.815
5 feet	0.605	16 feet	1.936
8 feet	0.968	18 feet	2.178
10 feet	1.21	20 feet	2.42
12 feet	1.452	24 feet	2.904
14 feet	1.694	25 feet	3.025

MILES TRAVELED IN PLANTING AN ACRE—3′ 6″ ROWS

1-Row Planter	2.34 miles
2-Row Planter	1.17 miles
3-Row Planter	.78 miles

ACRES PLANTED IN TRAVELING ONE MILE—3′ 6″ ROWS

1-Row Planter	.42 acres
2-Row Planter	.84 acres
3-Row Planter	1.26 acres

There are 10,667 stalks in an acre planted in 3′ 6″ rows, three stalks to the hill, hills 3′ 6″ apart, or drilled one stalk every 14 inches.

There are 3,556 hills in an acre planted in 3′ 6″ rows, hills 3′ 6″ apart.

U. S. GOVERNMENT LAND MEASURE

A township—36 sections, each a mile square.

A section—640 acres.

A quarter section—half a mile square, 160 acres.

An eighth section, half a mile long, north and south, and a quarter of a mile wide—80 acres.

A sixteenth section, a quarter of a mile square—40 acres.

The sections are all numbered 1 to 36, commencing at the northeast corner.

The sections are divided into quarters, which are named by the cardinal points. The quarters are divided in the same way. The description of a forty-acre lot would read: The south half of the west half of the southwest quarter of section 1 in township 24, north of range 7 west, or as the case might be, and sometimes will fall short and sometimes overrun the number of acres it is supposed to contain.

NOTE—In most of the western states, where all of the land was laid out by the Government, all titles, except in city lots, are passed by description, as under the Government survey, and there a square of 6 miles, or 36 square miles, is one township.

LAND MEASURE

To find the number of acres in a body of land, multiply the length by the width (in rods) and divide the product by 160. When the opposite sides are unequal, add them, and take half the sum for the mean length or width.

TO MEASURE CORN IN CRIBS

Corn in the ear of good quality, measured when settled, will hold out at 2½ cubic feet to the bushel. Inferior quality, 2¼ to 2½ cubic feet.

Rule—At 2⅜ cubic feet to the bushel, divide the cubic feet in crib by 2⅜, or multiply by 8 and divide by 19.

At 2½ cubic feet to the bushel, divide the cubic feet in crib by 2½, or multiply by 2 and divide by 5.

TO FIND THE NUMBER OF TONS OF HAY IN A MOW

Multiply the length by the width by the height (all in feet) and divide by 400 to 500 depending on the kind of hay and how long it has been in the mow.

TO FIND THE NUMBER OF TONS OF HAY IN A STACK

Multiply the overthrow (the distance from the ground on one side over the top of the stack to the ground on the other side) by the length by the width (all in feet); multiply by 3; divide by 10, and then divide by 500 to 600, depending upon the length of time the hay has been in the stack.

CAPACITY OF CORN CRIBS. (Dry Corn.)

Height, 10 Feet.

Length	½	1	12	14	16	18	20	22	24	28	32	36	48	64
Width 6	13	27	320	373	427	480	533	587	640	747	853	960	1280	1707
6¼	13	28	333	389	444	500	556	611	667	778	889	1000	1333	1777
6½	14	29	347	404	462	520	578	636	693	809	924	1040	1387	1849
6¾	15	30	360	420	480	540	600	660	720	840	960	1080	1440	1920
7	16	31	373	436	498	560	622	684	747	871	996	1120	1493	1991
7¼	16	32	387	451	516	580	644	709	773	902	1031	1160	1547	2062
7½	17	33	400	467	533	600	667	733	800	933	1067	1200	1600	2133
7¾	17	34	413	482	551	620	689	758	827	964	1102	1240	1653	2204
8	18	36	427	498	569	640	711	782	853	996	1138	1280	1707	2276
8½	19	38	453	529	604	680	756	831	907	1058	1209	1360	1813	2418
9	20	40	480	560	640	720	800	880	960	1120	1280	1440	1920	2560
10	22	44	533	622	711	800	889	978	1067	1244	1422	1600	2133	2844

The length is found in top line, the width in left-hand column—the height being taken at 10 ft. Thus, a crib 24 ft. long, 7½ ft. wide and 10 ft. high, will hold 800 bushels of ear corn, reckoning 2½ cubic feet to hold a bushel. If not 10 ft. high, multiply by the given height and cut off right-hand figure. If above crib were only 7 ft. high, it would hold 800 x 7, equals 560 (0 bu., etc.). The same space will hold 1-4/5 times as much grain as ear corn. Thus, a crib that holds 800 bushels of ear corn will hold 800 x 1-4/5, or 1440 bushels of grain.

CAPACITY OF SILO

A silo, properly filled—that is, if the contents are made compact throughout—contains one ton of silage for every fifty cubic feet of space. To illustrate the economy of a silo to store stock feed as compared with a barn, a ton of hay requires 400 cubic feet of space. A farmer can easily figure how much a silo will contain by the following rules:

Multiply the square of the diameter by 0.7854, which will be the area of the circular floor. Multiply the area of the floor by the height, which will give the number of cubic feet. One cubic foot of silage weighs 40 lbs. Multiply the cubic feet by 40, and the result is the number of pounds of silage the silo will contain. Divide that by 2,000 to find the number of tons.

Diameter	Depth	Capacity in Tons	Acres to Fill 15 Tons to Acre	Cows It Will Keep 6 Months 40 Lbs. per Day
10	20	31	2-1/3	8
12	20	45	3	12
12	24	54	3-3/5	15
12	28	63	4-1/5	17
14	22	67	4-1/2	18
14	24	74	5	20
14	28	87	5-2/3	24
14	30	93	6	26
16	24	96	6-2/5	27
16	26	104	7	29
16	30	120	8	33
18	30	152	10-1/5	42
18	36	183	12-1/3	50

CISTERN CAPACITY

A cistern ten feet in diameter and nine feet deep will hold 168 barrels.

A cistern five feet in diameter will hold five and two-thirds barrels for every foot in depth.

A cistern six feet in diameter will hold six and three-fourths barrels for every foot in depth.

A cistern eight feet in diameter will hold nearly twelve barrels for every foot in depth.

A cistern nine feet in diameter will hold fifteen and one-half barrels for every foot in depth.

A cistern ten feet in diameter will hold eighteen and three-eighths barrels for every foot in depth.

TO FIND THE CONTENTS OF SQUARE TANKS IN GALLONS

Rule—Multiply the area of the bottom by the height in order to secure the cubic feet. Multiply the cubic feet by 7½ (exact 7.48) and the result will be the number of gallons. For the contents in barrels, multiply the cubic feet by .2375.

TO FIND THE VALUE OF ARTICLES SOLD BY THE TON

Multiply the number of pounds by the price per ton, point off three places and divide by 2.

TO FIND THE CONTENTS OF BARRELS AND CASKS IN GALLONS

Rule—Multiply the square of the mean diameter (in inches) by the depth (in inches) and the product by .0034.

CIRCLES AND GLOBES

To find the circumference of a circle, multiply the diameter by 3.1416.
To find the area of a circle, multiply the square of the diameter by .7854.
To find the surface of a globe, multiply the square of the diameter by 3.1416.
To find the solidity of a globe, multiply the cube of the diameter by .5236.

COMMODITY WEIGHTS AND MEASURES

A pint's a pound—or very nearly—of the following: Water, wheat, butter, sugar blackberries.

A gallon of milk weighs 8.6 pounds; cream, 8.4 pounds; 46½ quarts of milk weigh 100 pounds.

A keg of nails weighs 100 pounds. A barrel of flour weighs 196 pounds; of salt, 280 pounds; of beef, fish or pork, 200 pounds; cement (4 bags) 376 pounds.

Cotton in a standard bale weighs 480 pounds. A bushel of coal weighs 80 pounds.

A barrel of cement contains 3.8 cubic feet; of oil, 42 gallons.

A barrel of dry commodities contains 7.056 cubic inches, or 105 dry quarts.

A bushel, leveled, contains, 2,150.42 cubic inches; a bushel heaped—2,747.7 cubic inches. (Used to measure apples, potatoes, shelled corn in bin.)

A peck contains 537.605 cubic inches. A dry quart contains 67.201 cubic inches.

A board foot = 144 cubic inches; a cord contains 128 cubic feet.

WEIGHTS AND VOLUMES OF WATER

One cubic inch of water weighs .036 pounds. One cubic foot weighs 62.5 pounds. One cubic foot = 7.48 gallons. One pint (liquid) weighs 1.04 pounds. One gallon weighs 8.355 pounds. One gallon = 231 cubic inches. One liquid quart = 57.75 cubic inches.

TABLES CONVENIENT FOR TAKING INSIDE DIMENSIONS

A box 24 x 24 x 14.7 inches will hold a barrel of 31½ gallons.
A box 15 x 14 x 11 inches will hold 10 gallons.
A box 8½ x 7 x 4 inches will hold a gallon.
A box 4 x 4 x 3.6 inches will hold a quart.
A box 16 x 12 x 11.2 inches will hold a bushel.
A box 12 x 11.2 x 8 inches will hold a half-bushel.
A box 7 x 6.4 x 12 inches will hold a peck.
A box 8.4 x 8 x 4 inches will hold a peck, or four dry quarts.
A box 6 x 5.6 x 4 inches deep will hold a half-gallon.

TO FIND HEIGHT OF TREE OR BUILDING

Set up a stick and measure its shadow. Measure length of shadow of tree. Length of shadow of tree, times height of stick divided by length of shadow of stick equals height of tree.

COMMON MEASURES

Long Measure

12	Inches	1 Foot
3	Feet	1 Yard
5½	Yards	1 Rod
320	Rods	1 Mile
1	Mile	5280 Feet

The following are also used:

1 Size...........................1/3-Inch

(Used by shoemakers.)

1 Hand.........................4 Inches

(Used in measuring the height of horses.)

Fathom.........................6 Feet

(Used in measuring depths at sea.)

1 Knot.........................1.15 Miles

(Used in measuring distances at sea.)

Square Measure

144	Square Ins.	1 Square Ft.
9	Square Ft.	1 Square Yd.
30¼	Square Yds.	1 Square Rd.
160	Square Rods	1 Acre
640	Acres	1 Square M.

An acre is equal to a square whose sides are 208.71 feet.

Surveyor's Square Measure

10,000	Square Links	1 Sq. Chain
10	Square Chains	1 Acre
10	Chains Square	10 Acres

Surveyor's Linear Measure

7.92	Inches	1 Link
100	Links	1 Chain
80	Chains	1 Mile

Gunter's Chain is the unit and is 66 feet long.

Dry Measure

2	Pints	1 Quart
8	Quarts	1 Peck
4	Pecks	1 Bushel

1 Bushel contains 2150.42 cubic inches or approximately 1¼ cubic feet.

Liquid Measure

4	Gills	1 Pint
2	Pints	1 Quart
4	Quarts	1 Gallon

1 Gallon contains 231 cubic inches.
1 Cubic Ft. equals 7½ gallons.

Cubic Measure

1728	Cubic Inches	1 Cubic Ft.
27	Cubic Feet	1 Cubic Yd.
128	Cubic Feet	1 Cord

SUITABLE DISTANCES FOR PLANTING

Apples—Standard	30 to 40 feet apart each way					
Pears—Standard	16	" 20	"	"	"	"
Pears—Dwarf		10	"	"	"	"
Cherries—Standard	18	" 20	"	"	"	"
Plums—Standard	16	" 20	"	"	"	"
Peaches	16	" 18	"	"	"	"
Apricots	16	" 18	"	"	"	"
Currants	3	" 4	"	"	"	"
Gooseberries	3	" 4	"	"	"	"
Raspberries	3	" 5	"	"	"	"
Grapes	8	" 12	"	"	"	"

To estimate the number of plants required for an acre, at any given distance, multiply the distance between the rows by the distance between the plants, which will give the number of square feet allotted to each plant, and divide the number of square feet in an acre (43,560) by this number. The quotient will be the number of plants required.

NUMBER OF POUNDS TO THE BUSHEL

Alfalfa	60	Kafir Corn	56
Barley	48	Lime	80
Beans (White)	60	Malt	38
Bran	20	Millet Seed, Common	50
Buckwheat	48	Oats	32
Blue-Grass Seed	14	Onions	57
Clover Seed	60	Orchard Grass	14
Clover (Sweet)	60	Peas	60
Corn (Shelled)	56	Potatoes	60
Corn (In Ear)	70	Red Top Seed	14
Corn (Kafir)	56	Rye	56
Coal, Hard	80	Timothy Seed	45
Hubam Seed	60	Wheat	60
Hungarian Grass Seed	45		

LUMBER MEASURE

To find the contents of boards, in square feet.   Rule—Multiply the length (in feet) by the width (in inches) and divide the product by 12.

   Example—Find the contents of a 16-foot board, 9 inches wide.
   9 x 16 = 144 ÷ 12 = 12 square feet.

   To find the contents of scantlings, joists, etc., in square feet.
Rule—Multiply the length, thickness, and width together, and divide the product by 12.

   Example—Find the contents of an 18-foot joist, 2 x 8.
   2 x 8 x 18 = 288 ÷ 12 = 24 square feet.

TO FIND NUMBER OF BOARD FEET IN A LOG

Subtract 4 inches from the diameter and square the remainder.   The result will be the number of board feet in a 16-foot log.   Add ⅛ for 18-foot logs, ¼ for 20-foot logs.   Subtract ⅛ for 14-foot logs, ¼ for 12-foot logs.

MISCELLANEOUS INFORMATION

A gallon of water equals 231 cubic inches and weighs 8-1/3 pounds.   A cubic foot of water equals 7½ gallons and weighs 62½ pounds.

Water expands 1/11 of its bulk in freezing.

One cubic inch of water evaporates into a cubic foot of steam.   To evaporate one cubic foot of water requires the consumption of 7½ pounds of coal, or about one pound of coal to a gallon of water.   Each nominal horse power of a boiler requires 30 to 35 pounds of water per hour.

One-inch of rainfall means 100 tons of water on every acre.

A column of water 2-3/10 feet high equals one pound per square inch pressure.   To find the pressure per square inch of a column of water, multiply the height of the column in feet by the decimal .434.

Doubling the diameter of a pipe or cylindrical vessel increases its capacity four times.

Double-riveting is from 16 to 20 per cent stronger than single-riveting.

To find the circumference of a circle, multiply the diameter by 3.1416.

To find the diameter of a circle, multiply the circumference by .31831.

To find the area of a circle, multiply the square of the diameter by .7854.

To find the sides of an equal square, multiply the diameter by .8862.

To find the capacity of cylindrical tanks, square the diameter in inches, multiply by the height in inches, and this product by the decimal .34.   Point off four decimals and you have the capacity in gallons.

To find the contents of a pile of cordwood, multiply the length, width, and height together and divide the product by 128.   This will give you the number of cords.

AMOUNT OF PAINT REQUIRED FOR A GIVEN SURFACE

It is impossible to give a rule that will apply in all cases, as the amount varies with the kind and thickness of the paint, the kind of wood or other material to which it is applied, the age of the surface, etc.   The following is an approximate rule: Divide the number of square feet of surface by 200.   The result will be the number of gallons of liquid paint required to give two coats; or divide by 18 and the result will be the number of pounds of pure ground white lead required to give three coats.

GESTATION TABLE

Date of Service		Date Animal Due to Give Birth							
		Mare		Cow		Ewe		Sow	
Jan.	1	Dec.	6	Oct.	10	May	30	April	22
Feb.	1	Jan.	6	Nov.	10	June	30	May	23
March	1	Feb.	3	Dec.	8	July	30	June	22
April	1	March	6	Jan.	8	Aug.	28	July	21
May	1	April	5	Feb.	7	Sept.	27	Aug.	20
June	1	May	6	March	10	Oct.	28	Sept.	20
July	1	June	5	April	9	Nov.	27	Oct.	20
Aug.	1	July	6	May	10	Dec.	27	Nov.	20
Sept.	1	Aug.	6	June	10	Jan.	26	Dec.	21
Oct.	1	Sept.	5	July	10	Feb.	25	Jan.	20
Nov.	1	Oct.	6	Aug.	10	March	27	Feb.	20
Dec.	1	Nov.	5	Sept.	9	April	26	March	22

Belting Pointers

HOW TO FIND LENGTH REQUIRED

When it is not convenient to measure with the tapeline the length required, apply the following rule: Add the diameter of the two pulleys together, divide the result by 2, and multiply the quotient by 3-1/4; then add this product to twice the distance between the centers of the shafts, and you have the length required.

If possible to avoid it, connected shafts should never be placed one directly over the other, as in such case the belt must be kept very tight to do the work.

It is desirable that the angle of the belt with the floor should not exceed 45 degrees. It is also desirable to locate the shafting and machinery so that belts should run from each shaft in opposite directions, as this arrangement will relieve the bearings from the friction that would result when the belts all pull one way on the shaft.

TO FIND THE BELT SPEED IN FEET PER MINUTE

Multiply diameter of pulley (in inches) by 3.1416. This gives circumference of pulley and this result multiplied by number of revolutions will give you belt speed in inches.

RELATIVE TRANSMISSION OF HORSE POWER FOR ANY GIVEN WIDTH OF BELT

The horse power for a given speed will be directly proportioned to the width of the belt; that is, a 4-ply, 16 inches wide, running at a certain speed, will transmit eight times as much power as a 4-ply belt, 2 inches wide, running at the same speed; and a belt 100 inches wide, ten times as much as a 10-inch belt of the same thickness, running at the same speed, etc.

TO FIND THE HORSE POWER THAT ANY GIVEN BELT WILL TRANSMIT ECONOMICALLY

Multiply the width of the belt in inches by its speed in feet and divide the result by 800. The final result will be the horse power for a 4-ply belt. For a 6-ply belt, divide this result by 600; for an 8-ply, divide by 400; for a 10-ply, divide by 350.

TO FIND THE PLY OF A BELT OF A GIVEN WIDTH REQUIRED

To transmit a given horse power, economically, at a given belt speed, multiply the given horse power by 800 and the given width in inches by the given belt speed in feet and divide the first result by the second.

If the final result is one, or nearly one, a 4-ply belt is required; if one and one-half, a 6-ply; if one and three-quarters to two, an 8-ply; if two to two and one-quarter, a 10-ply.

TO FIND WIDTH OF BELT REQUIRED

To find the width of a 4-ply belt required to transmit a given horse power at a given belt speed per minute: Multiply the given horse power by 800, and divide the results by the given belt speed.

To find the width of a 6-ply belt required: Multiply horse power by 600; divide result by belt speed.

To find the width of an 8-ply belt required: Multiply horse power by 400; divide result by belt speed.

To find the width of a 10-ply belt required: Multiply horse power by 350; divide result by belt speed.

TO FIND SPEED AND DIAMETER OF PULLEYS

The product of the diameter and speed of the driving pulley equals the product of the diameter and speed of the driven pulley; consequently, if the speed and the diameter of the driving pulley are given, multiply them together and divide by the diameter of the driven pulley to find the speed of the driven; or divide by the speed of the driven pulley to find its diameter.

Example—The drive pulley on a tractor is 9½ inches in diameter and runs at 1,000 R. P. M.; what size pulley must be used on a thresher cylinder shaft that must run 1,100 R. P. M.?

9½ times 1,000 equals 9,500; divided by 1,100, equals 8.64. Since pulleys are made only in certain standard diameters, use either the next size larger, 9-inch diameter, and raise the engine speed slightly, or use 8½-inch pulley, considering that the slight slippage will reduce the effective speed to the correct number of revolutions per minute.

Example—At what speed will a rock crusher run, if its 6-inch pulley is belted to a 9½-inch pulley on a tractor with a R. P. M. of 1,000?

9½ times 1,000 equals 9,500; divided by 6 equals 1,583 R. P. M.